AI UNLEASHED: TRANSFORMING MARITIME SHIPPING FOR THE FUTURE

By: Mustafa Nejem

PREFACE

Welcome to a new era of disruptive technologies and data-driven innovations in maritime shipping. The maritime world connects nations through trade routes and enables a seamless flow of commerce and trade. For the industry, a quiet revolution has suddenly penetrated all its business processes and modes of operations. Artificial Intelligence, abbreviated as AI, was once limited to the interests of computer science students. However, nowadays, its applications can be seen in all industries and sectors including the maritime shipping. The ripples of the AI-powered solutions are spreading across the globe.

I am highly enthusiastic and excited as the author of this book to take you on a journey of exploration where there will be amalgamation of AI-powered solutions and maritime transportation. This book is not just a guide on AI technology nor is the target readers exclusively technology professionals. Instead, this book highlights the profound transformation that AI has introduced to the maritime shipping industry and it will interest all the stakeholders including ship owners, ship managers, maritime institutes, and organizations.

Maritime shipping, also known as maritime transport, refers to the transport of goods, cargos, and passengers through the ocean route or waterways. It has also been mentioned in the academic literature as freight transport and the transportation by sea has been one of the oldest mode of transportation in the human history. Maritime shipping operations are executed with the help of intricate logistics, colossal vessels, and complex dynamics of the world trade. The connectivity of the maritime shipping operations with the algorithmic decision-making and data-driven solutions may appear worlds apart. However, throughout your journey in going through different pages and chapters of this book, you will come to know how the fusion of AI and maritime shipping holds a promising future for the industry.

AI is not just a technological jargon but a real-world force that has already penetrated into many industries and sectors. Organizations not embracing the AI tools are lagging behind in the industry. The future of maritime shipping will be highly dominated by AI tools and technologies and AI concepts will determine how goods should be transported and how ships should be navigated for the optimized routes. AI will also play a pivotal role in ensuring the safety of the sea travel, improving the fuel efficiency, and reducing the carbon emissions.

I got the opportunity of meeting with pioneers and experts in the maritime shipping and AI implementation. Their experiences and perspectives have helped me a lot in enriching the content of this book. The extensive research from the recent literature combined with the voices of the industry experts builds the core foundation of this book. As we embark on the journey of AI in maritime shipping, I encourage you to read every concept and chapter of this book with a sense of curiosity and an open mind. This book is written to be read from cover to cover. The chapters are connected to the concepts of the previous chapters and you will get the best of this book by reading it in a chronological order. AI in maritime shipping is a promising but complex initiative. This book will provide you guiding principles based on successful models and case studies, and it can be a source of inspiration for you.

I hope that the book truly inspire you to embrace the AI technologies for transforming the maritime shipping. I am grateful to you for being part of this journey on unleashing the potential of AI.

[Name of the Author]

[Date]

TABLE OF CONTENTS

List of Figures

INTRODUCTION

1.1. THE BASIC IDEA OF ARTIFICIAL INTELLIGENCE (AI)

Artificial Intelligence (AI) has now become a buzz word in the digital world. If you don't understand the basic concept of AI, you might not get a sense of AI implementation when in fact there is an implementation. For example, the concepts of AI are used in IoT environments for the detection of intrusions and prevention of cyber-attacks. Various machine learning techniques are used for this purpose. If you are working in an industrial sector where AI technology is being used for intrusion detection and automated response, it will occur in such a seamless manner that you will not notice the implementation of AI if you are not aware of what it is. On the other hand, some industry players may adopt an unethical claim that they have used AI in their technologies, whereas in reality, it is not a case. Therefore, it is highly significant for you to have a basic idea of AI.

AI, at its core, is the simulation of the processes of human intelligence. The computer systems, machines, and machine learning algorithms attempt to simulate the responses of human intelligence in given situations and scenarios.[i] Some of the fields in which AI has shown promising results are the development of expert systems, speech recognition, natural language processing, machine vision, and chatbots such as ChatGPT and Google Bard. AI implementation requires a specialized IT infrastructure for writing the software code and training the algorithms of machine learning. AI program development is not proprietary to any single programming language. AI developers use different programming languages such as Python, Java, R, Julia, and C++. When the AI programs are developed, the algorithms first ingest large datasets that contain the labelled data for training purposes. The algorithms analyze the data for finding patterns and correlations. These patterns are then used by the machine learning algorithms for predicting the future state of a problem or flagging an alert for any suspicious activity in the network systems.

The understanding of various fundamental concepts concerning AI will further expand your knowledge as to why AI is so popular these days and how its power can be unleashed for transforming the future landscape of maritime shipping. The first key aspect is intelligence simulation. In the AI implementations, human intelligence is simulated at the machine level. It is due to this fact, the tasks that were once performed by the humans are now easily performed by the robots such as the care robots in the healthcare sector or the inventory management robots in the supply chain management. AI developers create computer algorithms and programs such that the computer systems get the ability of processing the large volume of information. The algorithms perform the tasks of learning from the datasets, reasoning, and rational decision-making.[ii] The ultimate objective of the AI developer is to simulate the cognitive functions of human brain and make adaptations in the decision-making approach similar to humans.

The second key aspect of AI is learning and adaptation. The traditional computer programs were static in the sense that they were programmed once to perform a task and their functionality was limited to the extent that the features were available at the time of deployment. However, the AI programs learn from the new information and adapt to the emerging circumstances. Machine learning is regarded as a major subset of artificial intelligence. Machine learning techniques improve the performance of the AI algorithms as the AI program gets exposure to more and more data. In the real-world too, a maritime shipping company's HR department will give more preference to those candidates who not only possess the essential knowledge and skills but also have a vast experience of the field. It is because the experience exposes the individuals to various situations and circumstances. These scenarios improve the decision-making ability of the individual and the person becomes more adaptable for the new circumstances. This power of learning and adaptation is one of the most differentiated aspects of the AI programs that set them apart from traditional computer algorithms.

AI programs, unlike the conventional computer programs, do not require explicit instructions for all possible scenarios. Instead, they use their learning from the datasets to make intelligent decisions. This unique ability of the AI programs also comes up with a limitation. Computer algorithms are eventually machine learning algorithms and their decision-making is based on the training dataset available to them. If there is a problem for which no data is available to the algorithm, the algorithm will still use its best judgment but the accuracy of the results may not be guaranteed in those cases. Due to these reasons, the AI developers place a strong emphasis on the training datasets. The more reliable and comprehensive the training dataset is, the more reliable, accurate, and precise will be the results of the AI algorithms.

Another unique aspect of AI is problem-solving. The expert systems that are developed by using AI are good problem solvers. They will analyze a large volume of data and give to you the prompt results to complex problems that a human might take months to analyze and recommend. This power of AI is particularly relevant in maritime shipping where the human decision-making can take a lot of time. Even after spending a considerable time, the human processing is limited by the cognitive capabilities of the individual and is prone to error. One key dimension of AI is automation. The industrial-level AI systems are not just designed to make reports and identify errors. They are also designed to provide automated responses and take corrective actions. The tasks can be executed with minimal human intervention and the accuracy level of the execution also increases. Various industries such as banking sectors, healthcare sectors, and the transportation industry have implemented automation by the optimized use of AI-powered algorithms and robots. It has increased the efficiency of their business processes and streamlined their processes. It is high time that the same level of automation is also implemented in the maritime shipping. The AI algorithms follow a structured criteria for decision-making. Their decisions are based on predefined rules as well as their training from the dataset. They may also recommend a product based on their knowledge base or drive an autonomous vehicle by following the rules of driving. This ability of the AI algorithms to make data-driven decisions will reshape the future of maritime shipping.

A powerful aspect of AI implementations is the potential of sensing and perception. AI systems not only process the data efficiently but they can also interpret the information. They are highly responsive to the environmental variables and can easily be integrated with cameras, sensors, and other remote-sensing devices. This power of AI is further expanded by its ability to process auditory information and visual information. There are already technological solutions available in the market based on speech recognition, image recognition, and the processing of the natural language.

Another factor that helps in understanding the basic idea of AI is the power of AI algorithms to understand the natural language. For example, an AI care robot will not only give medications to the patients at the specified time with the prescribed doses but also respond to the requests of the patients.[iii] If a patient needs an emergency medication, s/he will 'talk' to the robot just like a human. The AI robot will understand the natural language and obey the instructions of the patient. AI systems have become highly responsive by understanding the human language. They can interact with users through the textual data as well as the speech data. The best examples in the modern day scenario for natural language processing are chatbots such as ChatGPT and Bard and the virtual assistants such as Siri.

With all these factors and dimensions emerging from the basic idea of AI, it is quite evident that AI can transform the future of maritime shipping. However, ethical and responsible use of AI is also critical and it is being emphasized in all industries and sectors. AI algorithms are developed by AI technology professionals and all subsequent processing is dependent on the development of these algorithms. If the developer is biased in the crafting of algorithms, inaccurate results may emerge that will misguide the organization and affect the overall productivity and profitability. Privacy concerns are also raised because AI algorithms need a large and relevant dataset for training purposes.[iv] Moreover, the users of AI systems often do not understand the internal working of AI and machine learning algorithms. Therefore, they have to trust the AI developers. If the developers are not men of integrity, confidential and sensitive data may reach unintended recipients.

The increased use of AI has also created a sense of insecurity among the currently hired workers in different industries. They believe that AI implementations will eat their jobs and more and more robots will replace the humans. The senior management in those organizations should make the workers realize that AI implementations are for improving the quality of output and enhancing the efficiency. The knowledge and experience of the hired workers will still be respected and the implementations will not result in downsizing or retrenchment.

Many individuals confuse the term machine learning and use it interchangeably with AI. As shown in

KEY COMPONENTS OF AI

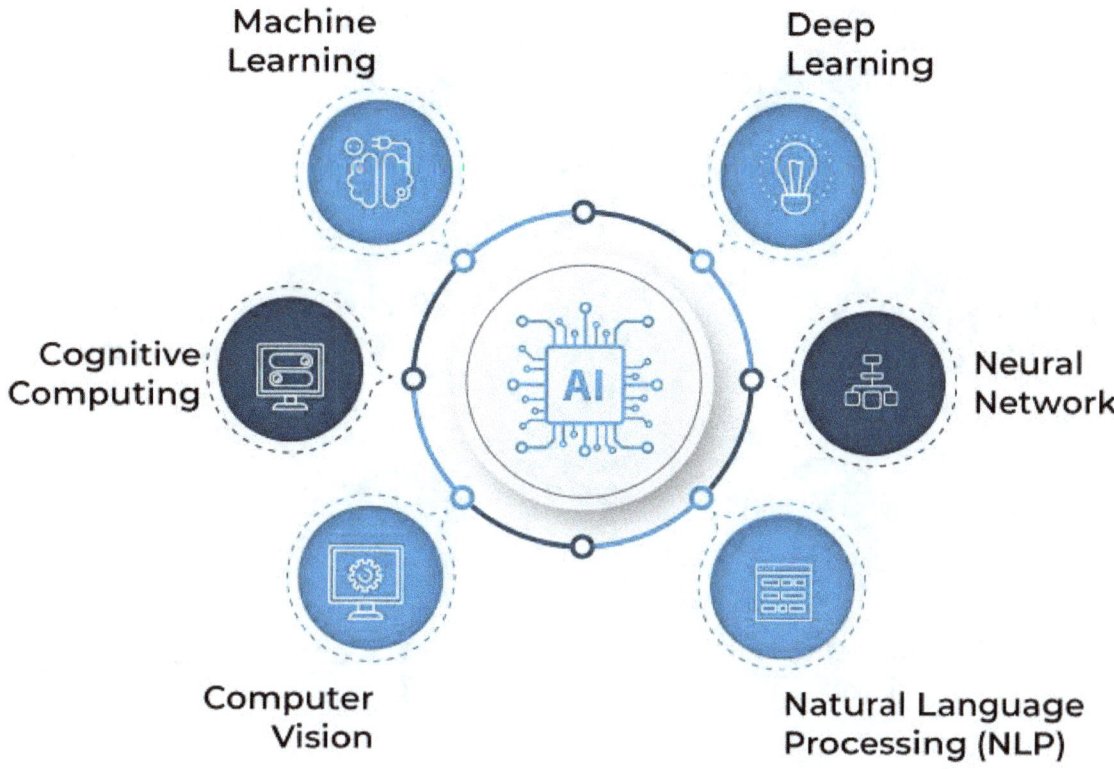

Figure 1 below, machine learning is just one of the key components of AI, otherwise AI is highly diversified domain. Other crucial components of AI include neural network, deep learning, natural language processing, computer vision, and cognitive computing.

KEY COMPONENTS OF AI

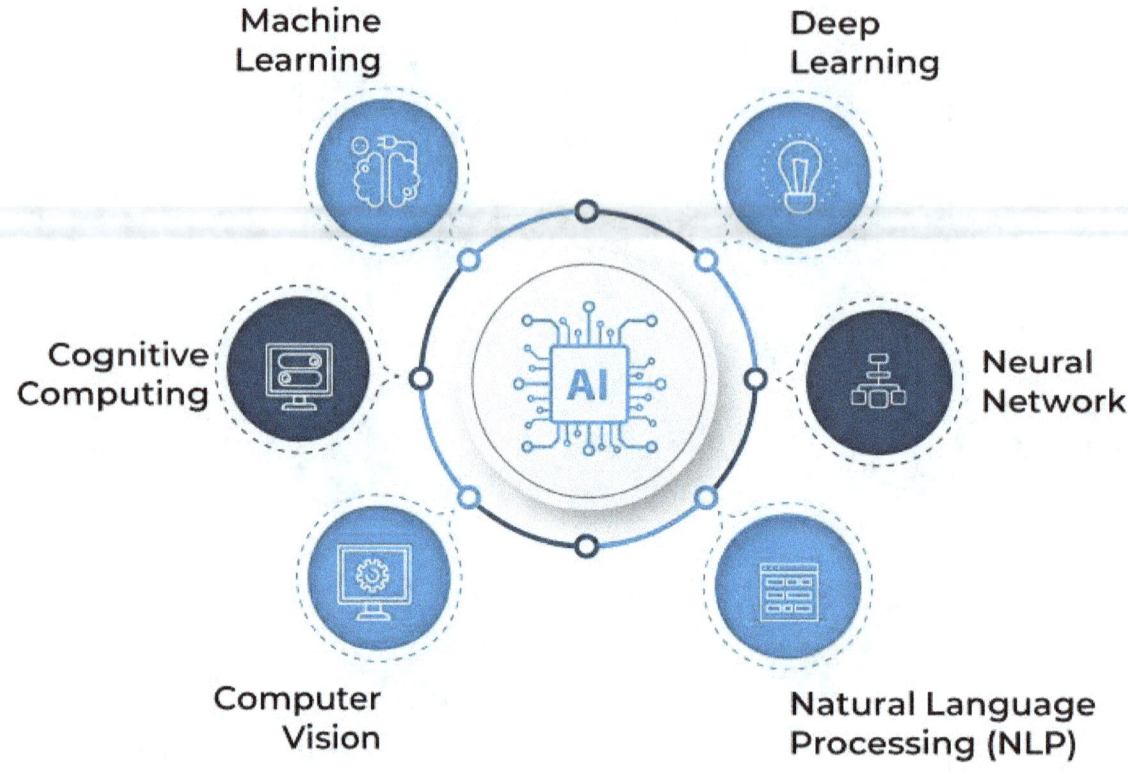

Figure 1: AI – Key Components[v]

AI – FROM CONCEPT TO APPLICATIONS

AI is not a new or recent concept. The course of artificial intelligence has been an integral part of computer science curriculum right from the beginning. The students of computer science used to learn AI concepts during their studies such as heuristic search, machine learning, Turing machine, theory of automata and formal languages, discrete mathematics, training datasets, and others. However, there was not a wide-scale implementation of these concepts because the industry leaders were not aware of the potential of AI.

Thanks to the academicians and scholars. Their scholarly writings in the current academic literature generated interest in AI implementation. When the AI expert systems were presented in a user-friendly way that even an average user was in a position to use these systems, the whole world geared to the AI technology. It was like an eye-opener. Just consider the recent AI-driven chatbots ChatGPT and Bard. They are as simple to use as if you are doing a WhatsApp chat to a real person. For a research on a particular topic, earlier you had to search a lot of articles on Google and then consolidate the findings to develop the write-up. But now all this work is being done by these chatbots. All you have to do is to write a 'prompt'. The 'responder' on the other side is not someone who is available to you for a limited time. You can talk to the responder 24/7 without any fear of a negative reply or punishment. During the classroom instructions, you might be a backbencher or fear asking questions with your professor, but now you have someone you can learn from as long as you want at any time of your convenience. This is the power of AI, from concept to application.

In the transportation sector, autonomous vehicles and driverless vehicles are gaining prominence that is all powered by AI. With the increased adoption of AI, a new term has emerged in the industrial sector known as Industry 4.0.[vi] In the Industry 4.0 paradigm, all devices and gadgets are interconnected and talk to each other. The maintenance issues are reported by the devices through sensors and automated actions may also be taken. As was put rightly by a wise person, 'we are living in a world, where you are no longer in need of right answers. All you are required is to ask the correct questions.' This is the power of AI. Just give the right training dataset to the AI algorithms and the rest will be taken care by the AI expert systems.

It is rather disheartening to note that maritime shipping has not embraced the AI technology with the same level of rigor and enthusiasm. Perhaps, there is a fear that the inaccurate results by AI systems will incur a massive loss to the management. But I would like to emphasize to ship owners, ship managers, maritime institutes and organizations that it is a matter of survival for them. If some of the companies take the lead in the AI adoption, then the non-adopters will be completely out from the competition. The adoption of AI improves the productivity and efficiency of the organization to such an extent that organizations with traditional modes will never be able to compete. The pioneering work of AI is credited to Alan Turing (

Figure **2**) that provided the basic framework for AI implementation.

FIGURE 2: ALAN TURING – PIONEER OF MODERN AI[vii]

He presented an ideal concept of a 'universal machine'. The machine could have performed any intellectual task because (theoretically speaking) it had infinite memory. The machine was supported with a simple head that could be used to move the machine from one state to another. The earlier challenges faced by the AI developers were regarding the availability of data and computing power. At that time, people were using mainframe computers in the organizations, and 386, 486, or Pentium computer machines for personal use as desktop computers. The data used to be stored on hard disks with limited storage and the data used to be carried on smaller disks known as floppy disks. It was difficult for the developers to simulate the thought processes of the humans because the limited computing power would not allow to do so. Even if it was accomplished in the testing environment, the programs would fail in the real-life, production environment on low-capacity desktop computers. Another complexity at that time was that it was an era of desktop applications. The web based applications were rare because the internet availability was very limited or absent. Therefore, any changes in the AI programs would necessitate making a new installation on the server machines and the client machines.

The late 20th century is regarded as a revolution in the AI concepts. At that time, high-speed, high-bandwidth internet was available to the masses and this era is also known as 'dot-com bubble'. The machine learning techniques gained a huge prominence in this era and the advanced concepts such as neural networks were used by the AI developers. The advanced systems made it possible to enable a good level of training and achieving adaptability based on the new interactions. The AI applications such as speech recognition, computer vision, and natural language processing are credited to this era in the computer world.

The next big development in the AI world was the rise of big data. The web 2.0 technologies such as social networking sites, blogs, and forums were adopted widely and user-generated content became a useful resource for the industry managers and directors. They wanted to improve their products and offer new products based on customer feedback and preferences. The rise of big data also enabled the availability of vast datasets for the training of AI algorithms. The rapid growth of AI was possible at that time because the machines with powerful configurations were being developed by hardware vendors. Machine learning techniques thrived with the rise of big data because the large volume of dataset is not a 'burden' for an AI algorithm but an opportunity of learning. The AI algorithms became more accurate and precise and it was possible to solve complex problems through these algorithms. There were practical applications for this paradigm shift in various domains and sectors including the maritime shipping sector.

In the contemporary context, the conceptual origins of AI have become quite mature and AI applications are making a profound impact in various industries and sectors. In the transportation sector, the most prominent application has emerged in the form of self-driving cars.[viii] These cars can transform the transportation industry because AI systems recommend the most optimized routes and reduce the fuel consumption. The cars are ridden by following all the rules of the traffic management systems. Another important sector embracing AI revolution is healthcare. We have seen care robots facilitating the task of nurses. The drug discovery, treatment plans, and medical diagnosis are all optimized by AI-powered systems. Machine learning techniques can even analyze the images and X-rays and predict epidemics and disease outbreaks.

AI algorithms are also being used in the financial sector for fraud detection as well as algorithmic trading. The algorithmic trading also enables the traders to engage in high-frequency trading and improve their profitability. The risk assessment and credit scoring are also being accomplished by AI applications. The banks have developed virtual assistants and chatbots to perform regular tasks such as balance inquiry, account details, and fund transfer. The manufacturing sector is utilizing AI in optimizing the quality control and manufacturing processes. The downtime of the sector has also been reduced because the predictive maintenance is used by the business managers for preventing the failures of equipment.

The retailers are also using AI-based recommendation algorithms in their e-commerce platforms. It has facilitated them to provide a personalized shopping experience to the customers. The efficiency has also been improved by using AI-based supply chain and inventory management systems. The educators are using AI-based learning platforms for presenting customized educational content to the students and promoting student-centered learning. AI tools have also improved the learning outcomes and student engagement. The entertainment industry, particularly the streaming services, is also using AI technology to recommend the most appropriate content to the users. Moreover, AI-generated art and music are gaining prominence. Now, it is possible through the AI tools to generate images and videos of real and imaginary persons. These artefacts are popular in the general public. In the blockchain world, crypto art or NFTs are also making use of the AI technology and earning huge revenues.

The energy sector also has an increased adoption of AI technologies, and it is also relevant for maritime shipping because the sea routes should also have an optimized travelling with minimal fuel consumption and carbon emissions. AI technologies and algorithms assist in the optimization of energy consumption. They can predict the failures of equipment in power plants and enable a seamless integration of renewable energy. One of the most useful and beneficial implementation of AI is in the security domain. AI tools and technologies facilitate the process of facial recognition and flag an alert for unknown faces. The expert systems can help in intrusion and threat detection. The systems can also provide protection against cyber frauds.

All these applications of AI tools and technologies highlight that there is a wide level of acknowledgment of the usefulness of AI. All industries should benefit from this emerging technology and promote a responsible and ethical use of AI.

THE INDUSTRIAL IMPLEMENTATION OF AI

The industrial implementation of AI can be understood in the broader context of Industry 4.0. Various technologies are considered as the driving forces in the Industry 4.0 paradigm and AI and machine learning is one of those forces.[ix] As shown in Figure 3 below, Industry 4.0 (I4.0) is often correlated with smart manufacturing because the focus in this approach is not on hard work and intensive labor but on executing the tasks smartly through digital revolution. I4.0 improves the productivity and enables real-time decision making. The whole factory works through the systems characterized by agility and flexibility.

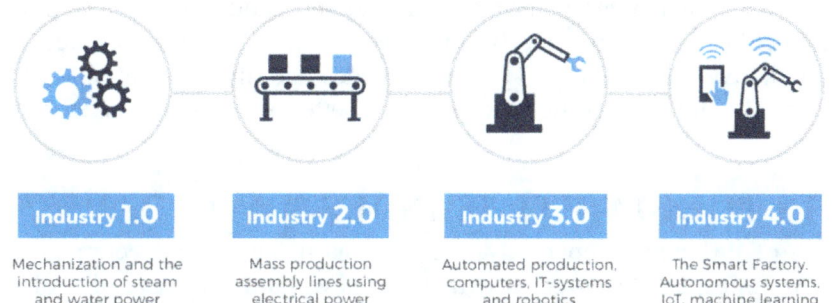

The Four Industrial Revolutions

Industry 1.0	Industry 2.0	Industry 3.0	Industry 4.0
Mechanization and the introduction of steam and water power	Mass production assembly lines using electrical power	Automated production, computers, IT-systems and robotics	The Smart Factory. Autonomous systems, IoT, machine learning

FIGURE 3: FROM INDUSTRY 1.0 TO INDUSTRY 4.0x

In the concept of smart factories, all the equipment used in the factory setting is equipped with the advanced sensors. Through the use of embedded AI software and the implementation of robotics, the data is collected and analyzed for the optimization of decision-making process. The value-creation may further be strengthened by the use of ERP systems, CRM, and supply chain modules. It provides a whole new level of insights and visibility to the critical manufacturing processes.

The I4.0 paradigm is termed as an integral component of the fourth industrial revolution. The first revolution began in the late 18th century in Britain. In that revolution, the mass production was enabled by utilizing steam and water power. It was termed as a revolutionary approach because the reliance was reduced on animal power and human power. Machines were also used for developing finished goods instead of relying solely on human hands for all stages of production. The second revolution is widely known for the increased adoption of assembly lines in the manufacturing process. At that time, the power sources were mainly electric, oil, and gas. The new power sources increased the wealth and economic prosperity of oil-producing countries. The advanced means of communication were also introduced such as telegraph and telephone.

The third revolution started in the middle of the 20th century when advanced digital and technical tools were introduced. It was a time when the world started the use of computers. Data analysis and advanced telecommunication technologies began to provide competitive advantages to the industrial sector. In the current era, from the industrial perspective, it is said that we are living in the fourth industrial revolution or I4.0. There is an increased demand by the industrial and business organizations for the automation of business processes. Smart factories, smart machines, and smart devices are the new talk of the town. The business managers believe in intelligent and data-driven decision-making. Products and service are being offered more efficiently throughout the value chain. Flexibility and agility are being embraced by the business organizations and customer demands are being met through mass customization. There is an increasing tendency to collect more and more data for the computer algorithms to generate intelligent insights. A smart data analytics enables the organizations to accomplish better decision-making, transparency, and accuracy.

When we talk specifically about I4.0, several technological innovations are driving this innovation, and AI is at the forefront of these innovations. These innovations include AI, IoT (Internet of Things), cloud computing, cybersecurity, edge computing, and digital twin.[xi] Through AI and machine learning techniques, it has become possible for the business entities to gain full benefits of the volume of data that is generated each day. It might be the transactional data or it may also be the user generated content. The organizations are also making an efficient use of data that is generated by third party and business partners. AI can provide insights and more visibility to the data. The data may also be used for making predictions and automating the business operations. As an example, industrial level machines are used in the maritime shipping operations. These machines are prone to damage and faults during the shipment process. When AI expert systems collect data from the business assets, they can help the decision makers to perform the tasks of predictive maintenance. This maintenance activity is accomplished by using machine learning techniques. The outcome is higher efficiency, better reputation, and enhanced uptime.

RELEVANCE OF AI IN THE MARITIME INDUSTRY TODAY

As I wrote at the very beginning of this book, apparently, there seems to be no connection of the maritime shipping industry and AI concepts. However, now that you have gained a basic idea of AI, you can very well appreciate that AI is extremely relevant for transforming the future of maritime shipping. Figure 4 below shows the immense opportunities and the potential of AI to transform maritime shipping. The world of today is a data-driven and fast-paced world. Therefore, the significance of AI should not be underestimated by the maritime industry professionals. Some maritime organizations have already embraced AI and it is eventually going to transform the maritime sector. I am convinced that AI will address the age-old challenges of maritime shipping and build a more sustainable, efficient, and competitive future. I have explained below various aspects and dimensions due to which I feel that AI is highly relevant for maritime shipping.

FIGURE 4: AI IN MARITIME SHIPPING OPERATIONSxii

The first area of consideration is the safety and navigation of the shipping vessel. The safety of the vessels is of paramount importance for ship owners, ship managers, maritime institutes and organizations. When AI-powered systems are used in the vessel operations, the navigation accuracy and precision of the vessel is improved.[xiii] AI developers have also introduced collision avoidance algorithms that further improve the safety of vessels. These AI systems possess the capability of processing large volumes of data. They receive data from weather forecast stations, sensors, and historical navigation records. It enables the captain of the vessel to make real-time intelligent decisions and avoid the risk of collisions and accidents. AI systems also empower the vessel staff to perform predictive maintenance. It assists them in identifying the possibility of equipment failures. Therefore, there is a reduced risk of accidents and breakdowns when the vessel is present at sea.

Another crucial aspect is that AI systems improve the operational efficiency of the vessels. The systems optimize the fuel consumption and vessel routing by analyzing different data parameters such as traffic patterns, weather data, and sea currents. When the ship routes are optimized, there is a reduced level of carbon emissions, fuel costs, and shorter durations in transit. The key business processes in maritime shipping are loading processes and cargo handling. These processes can also be streamlined and made efficient by the use of AI-driven robots. The decision-making process can also be optimized by the power of AI. AI systems are also accompanied by decision support tools. These tools can be highly valuable for ship operators and ship captains for making critical choices during their journey. As AI tools generate results based on real-time data, the captains can utilize the information for fuel consumption, route optimization, and predicting weather patterns. A unique feature of machine learning techniques is that they not only rely on training data but also evaluate the historical data. It assists the algorithms in identifying trends and patterns and making data-driven decisions concerning cargo allocation, route planning, and scheduled maintenance.

A differentiated aspect of AI systems is the notion of environmental sustainability. The systems can make a significant contribution in the efforts of the maritime shipping industry to minimize their carbon footprint.[xiv] It can be accomplished by a reduced fuel consumption and an optimized selection of routes. AI-powered systems are supported with monitoring systems and sensors. As a result, there is a higher level compliance of the business entity with the environmental regulations of the territory. The systems are well aware of the local and global regulatory frameworks and ensure the adherence of the vessels to emission limits. The systems can even implement measures for protecting the marine life and marine ecosystem. As shown in Figure 5 below, AI-powered systems achieve a higher level of environmental efficiency. They evaluate the variables of both fuel price and the estimated revenue and create a data pair. Then a learning framework begins with the updated belief regarding the current environment. Then the environment estimates are incorporated into the system. Based on these estimates, an optimized framework is created by AI-powered systems. The working of AI-based systems does not stop here because the systems believe in the process of continued learning. Therefore, this loop keeps going and the system refines its accuracy, prediction capability, and precision level.

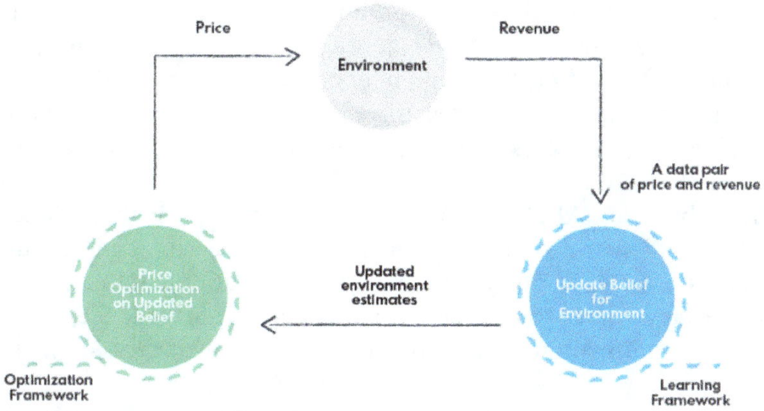

FIGURE 5: AI-BASED LEARNING AND OPTIMIZATION FRAMEWORK FOR

SUSTAINABILITY[xv]

Many maritime institutes and organizations are reluctant to implementing AI because they feel that it will incur a significant initial cost. There is no denying of the fact that the development of the required IT infrastructure and the acquisition of computer systems with high processing power will incur a significant initial cost. However, ship owners, ship managers, maritime institutes, and organizations should have a long-term orientation. They should analyze how it will make the life easier and improve the accuracy and safety of the whole maritime shipping operations. When predictive maintenance will be accomplished by AI-powered systems, the downtime will be reduced significantly and the maintenance costs will also decrease. When the equipment will be maintained regularly, the lifespan of the critical and expensive equipment will also increase in the context of maritime shipping. Furthermore, AI systems are also supported with AI analytics.[xvi] The use of the analytics feature will facilitate shipping companies in the identification of cost-saving opportunities. The cost savings can be achieved in supply chain management, vessel operations, and fuel consumption.

The loading and proper handling of cargo are considered critical operations in maritime shipping. AI-based systems can make use of sensors and blockchain technology. These innovations make it possible to track the cargo location in real-time and provide a secure documentation to the users and other stakeholders. The systems ensure the integrity of the entire supply chain. The security and safety of the goods can also be ensured during the transit. The systems almost eliminate the chances of theft or mishandling of the cargo.

Another important dimension is that AI-based systems are developed based on global standards, best practices, and authentic datasets. Therefore, the use of these systems can also enable the organizations to claim the compliance of their business processes with the international safety standards and regulatory frameworks. The shipping companies can save them from various penalties and punishments and maintain a good track record of their safety standards. Another reason due to which AI-powered systems are being preferred in various industries is their preference of providing data-driven insights. The systems are capable of processing large volumes of data and the data can be handled with multiple data sources and data types. Therefore, the shipping companies not only gain insights into the maritime shipping operations but also get valuable information regarding the market trends and customer demands. The strategic directions of the company can then be set based on these valuable insights.

1.2. POTENTIAL APPLICATIONS OF AI IN THE MARITIME INDUSTRY

Based on the maturity level of AI technologies in the current context, I can visualize various implementations based on AI systems that can prove highly beneficial for maritime shipping. Through this book, I am presenting a strong case for applying AI in the maritime shipping. The systems and processes of the maritime shipping industry are closely connected to other industries. Therefore, the industry cannot operate in isolation. Sooner or later, the industry professionals will be compelled to embrace the technological framework of AI tools and innovations.

The first potential application of AI is in unmanned operations and autonomous ships. It is a highly crucial application because the world is not a place of peace and love everywhere. We are facing continuous terrorist threats. Moreover, there are war-like situations in certain areas such as Ukraine and Yemen. When the shipping companies have to transport goods in those sensitive destinations, they risk the lives of their staff and crew members. Moreover, the cargo may not reach the desired location due to political unrest or prohibitions on travelling through sea route. In those cases, autonomous and unmanned ships can be highly valuable. In those cases, the worst case scenario is the loss of cargo, but still, the valuable human lives can be saved.

Autonomous ships can be constructed by using AI algorithms.[xvii] These ships are capable of operating and navigating independently in the sea. In the normal circumstances, these ships can reduce the operational costs and the safety features can also be optimized. In war-like situations, autonomous ships are also accompanied by unmanned aerial vehicles so that there is also an aerial surveillance of the area. The ship owners, ship managers, maritime institutes, and organizations can then enable emergency response and maritime security.

Figure 6 below shows the conceptual model of an autonomous ship. There is a ship-side activity powered by edge analytics and there is also a shore-side activity powered by cloud analytics. The ships use the intelligence of AI algorithms for context awareness, controlling the ship, interacting with the ship, and making judgments. The shore side is linked to the ship side through the satellite connection. The maintenance activity and monitoring are accomplished at the shore-side. Data integration and analysis are also performed at the shore side.

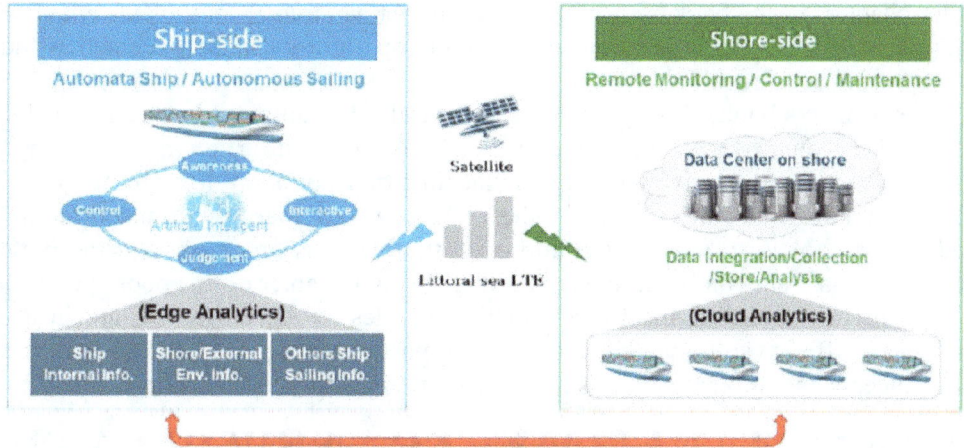

FIGURE 6: CONCEPTUAL FRAMEWORK OF AUTONOMOUS SHIPSxviii

Another useful application of AI in maritime shipping is concerning fuel efficiency and route optimization. The algorithms accomplish a dynamic weather forecasting and the vessels can make adjustments in the routes in real-time. As a result, the vessel operations are prevented from adverse weather conditions and a safe delivery of the cargo is ensured.

As shown in Figure 7 below, the ship routes are optimized by AI algorithms in a structured format. The first step in this case is the collection of relevant data. The data includes the map data as well as the weather data. Then machine learning algorithms perform a supervised learning and build weather predictions. The map projections are also taken into consideration. The final stage is the optimization of ship route based on the improved belief system. The AI algorithms build an adaptive grid system and show the optimized route in a graphical format. This route is based on the weather conditions as well as historical data of the safety aspects in a given area. The captains can then confidently follow the recommended route for fuel efficiency and improve safety.

FIGURE 7: AI-POWERED OPTIMIZATION OF SHIPPING ROUTESxix

Another potential application of AI is in the domain of cargo loading and cargo handling. The robotic systems can be highly relevant in this context. With the use of robotics, the loading and handling processes can be automated. It will improve the efficiency of the operations and the system will also operate effectively in challenging conditions. Computer vision is a powerful tool in AI. The algorithms can be used for sorting the cargo and the proper handling of the cargo can be ensured throughout the vessel operations.

Figure 8 below shows automated cranes that also possess the capabilities of object detection. These cranes can differentiate between objects and workers and the cargo can be transported from one place to another without collisions. The delays and accidents are avoided during the whole operations.

FIGURE 8: ROBOT CRANES IN MARITIME SHIPPINGxx

The examples above reveal that AI has a promising future for the maritime industry. I have listed only a few examples in the introduction section. As you will read other chapters of the book, you will realize that AI can transform the entire landscape of the maritime shipping. The tasks that are consuming numerous hours can be executed without human interventions by using AI algorithms.

In the first chapter of this book, I gave you a basic idea of AI and how it is relevant to the maritime transportation. When you read the next chapter of this book, you will further appreciate my arguments because now I am focusing exclusively to AI and maritime transportation. I am going to present to you successful case studies from the real-life where AI-powered maritime transportation has been implemented. I will explain to you how it improved the business processes and the bottom line of maritime organizations. I am confident that by reading the first chapter of this book, you have gained a good level of expertise regarding the foundational principles of AI. Now, you are in a position to evaluate and assess different solutions in the context of maritime shipping. Let's go to the next chapter and see what concepts and implementations have been highlighted in each section.

2. AI AND MARITIME TRANSPORTATION

Despite the advancements in air routes and drone technologies, maritime transportation still has its own significance. It can be regarded as the lifeblood of global commerce and trade. The moving of goods through sea routes is also connecting nations. The nations with good trade volumes also enjoy good political and social relationships. The maritime transportation industry needs to be revolutionized through the disruptive innovation of AI. This technological evolution has the potential of making a paradigm shift in the navigation, management, and operations of ships. The sustainability and environmental considerations are being advocated at the national and international levels. If there is one technological concept that can reshape the maritime shipping industry in this regard, it is the power of AI. The integration of AI into maritime shipping will unlock numerous possibilities for the industry concerning environmental sustainability, safety, and efficiency. A great aspect of AI technologies is that they continue to evolve and advance. Therefore, the maritime shipping sector should not only embrace the current AI technology but also be prepared for further innovations. This mode of continuous innovation will set a stage for a connected, technologically-advanced, and environmental-friendly future of the maritime shipping.

In the maritime sector, the ship crew and vessel operations can experience a variety of benefits by using AI. The data tracking and useful insights can reduce the cost of voyage significantly. The efficiency of the voyage can be improved regarding the fuel efficiency as well as route optimization. The shipping organizations can also contribute to the cause of environmental sustainability by reducing their carbon footprints. There is also a trend of automating the vessel operations by using AI-powered systems. The use of these systems will eliminate the risk of human error and ensure the safety of ship crew, cargo, and the machines. The sections below highlight the benefits of AI in maritime shipping and then present successful cases to make the readers realize that AI is in fact successful in maritime shipping. After presenting the case studies, I have elaborated the significance of AI in each individual operations including AI in vessels, predictive maintenance, cargo optimization, improved navigation and safety, fuel efficiency, route optimization, and prevention collisions.

1.1. BENEFITS OF AI IN IMPROVING MARITIME TRANSPORTATION

AI has been introduced in different industrial and service sectors as a tool of revolutionizing the business processes. The manner in which ships are operated, managed, and monitored today will be a thing of the past when AI technologies are integrated. I have mentioned below some of the key benefits of AI in optimizing and improving maritime transportation.

Based on the capability of AI technologies to simulate human behavior, these technologies have been categorized into three main categories. The first is known as ANI (Artificial Narrow Intelligence).[xxi] It has been named so because these algorithms work based on the preset limits defined by the AI developers. These systems are trained by a huge dataset. Its performance may even surpass human intelligence; however, its performance is limited to the capabilities that have been designed by the AI developers. An example in this regard is that of chess programs that are developed based on AI concepts. These programs can even beat the world champions of chess. However, if you expect these programs to perform any tasks other than playing chess, they will be helpless. It is due to this fact this category is considered as a 'narrow intelligence'.

The next category is known as AGI (Artificial General Intelligence).[xxii] In this stage, the machine becomes a learned device. These devices can also perform intellectual tasks by simulating the human behavior. This stage of AI is still in the development stage but it has also raised alarms among the technology and non-technology communities. It is being argued that whether the human community is prepared for welcoming those 'brains' that are superior to human brains. It is being feared that these artificial brains will result in increased joblessness and make the tasks of many professionals redundant. It is also being argued that the robots might take the control over cultures and civilizations.

The third category is known as ASI (Artificial Super Intelligence).[xxiii] It is a stage when artificial intelligence beats the human power and goes ahead in all operations and circumstances. When ASI technology is mature, it will be a real challenge for the humans and organizations because ASI-based systems will improve their intelligence in a very short time. A human being takes several years to complete his/her formal education. Similarly an organization takes numerous years to make its presence felt in the industry. However, ASI-based systems will be extensible in no time. It poses a challenge for the maritime institutes and organizations as well. If they do not integrate AI in their systems and procedures, they will soon be lagged behind and their business operations will not be sustainable. As shown in Figure 9 below, AI-based systems are classified into three categories based on their capabilities. The systems may also be categorized based on their functionalities. Reactive machines systems only respond to the stimuli and they do not have any proactivity in their approaches. The historical data and patterns are not crucial for this type of AI-based systems and they are only appropriate for game playing machines. Limited memory systems save data regarding past experiences and learn from the dataset. Therefore, they can recommend a calculated solution for a given problem.

The systems based on the Theory of Mind are still in the beginning stage and these systems can also evaluate human emotions, concepts, and beliefs. They also possess the capability of social interaction such as the care robots in the healthcare system. The self-aware systems, just like the Maslow's hierarchy where self-realization is the peak of the pyramid, are considered as the highest level of the AI development. It is a theoretical concept at present. Such an AI-based system will be a true copy of the human brain filled with emotions and sentiments. Such systems and robots might be seen in the future, but the world is yet to see a functional model of this theoretical concept.

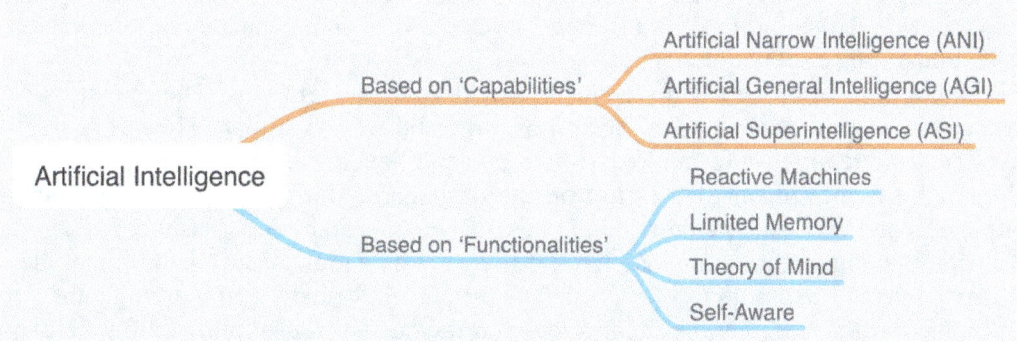

FIGURE 9: AI SYSTEMS CATEGORIZATIONxxiv

When it comes to the vessel operations, the safety and security of the journey is of paramount significance. Imagine a ship is stuck in the sea. How will it reach the destination? How much information is available to the control room regarding the ship, crew staff, and cargo? Why the control failed to prevent the occurrence of this incident? These are the questions that immediately come to the minds of ship owners, ship managers, maritime institutes and organizations. AI is all about making an efficient use of the available data and adopting a proactive strategy. The incidents are avoided at first, and if they inevitably occur, the response is so quick that there is minimal loss to the human lives, property, and belongings. When AI-powered predictive analytics is used, the risks and safety hazards are predicted by the AI algorithms. Therefore, it is possible for the ship managers to introduce proactive measures. A safer maritime environment benefits both the vessels and the crew. When collision avoidance systems are made by using AI technologies, there is a real-time monitoring of the entire ship operations and there is a safe journey of the vessel through complex routes and congested waters. As shown in Figure 10 below, AI-powered predictive analytics can make an efficient use of big data for improving the efficiency and reducing delays in maritime shipping. Valuable pieces of information also make it possible for ship managers to avoid risks, collisions, and accidents.

FIGURE 10: PREDICTIVE ANALYTICS AND BIG DATA PROCESSINGxxv

The use of big data can provide benefits to the maritime shipping in several areas. The first crucial area is chartering. Charterers have to find the right ship for the cargo so that the goods are delivered to the destination with the most efficient pricing. At present, maritime institutes and organizations rely heavily on the information provided by the ship owners and brokers. However, their information and advice might be limited and biased. AI-powered data analytics can be highly useful and charterers can base their decisions on accurate and actionable data. They can use different data sets such as position reports, vessel information, estimated time of arrival, market data, and Automatic Identification System (AIS). This information can be entered into an online portal of AI system to view all the alternatives available for the ship selection. The access to more information and alternatives increases the transparency and improves the competitiveness of ship operations.

Another area of consideration is the operational side of the vessel. Ships, similar to some other vehicles, have normal and optimum speeds. At any given point in time, there is an optimum speed of the ship that can provide maximum fuel saving. However, the ship captain finds it extremely challenging to operate the ship at the optimum speed. It is because several factors come into play to measure the optimum speed such as the maintenance level of the vessel and the engine wear. When big data analytics and predictive analytics are used, the ship captain can easily determine the optimum speed for the maximum fuel saving. In fact, it is shown as a dashboard indicator to the vessel's captain.

Vessels are heavy and highly expensive assets and there should be a scheduled maintenance of all the parts making up a vessel. Ship owners and ship managers have to make decisions as to when hull cleaning and propeller polishing needs to be carried out. In the absence of AI technologies, they base these decisions on intuition or a set schedule without considering the actual performance of the vessel. AI-powered algorithms, on the other hand, critically analyze the data of fuel consumption. They perform a cost-benefit analysis and then recommend a customized schedule for vessel maintenance. Data analytics facilitates the ship operators to know the benefits of maintenance and understand the most appropriate timing of vessel maintenance.

AI-powered technologies also prove highly useful in actual voyage operations. The operators of the terminal and port agents need to have a precise detail of cargo and ETA (Estimate Time of Arrival). AI algorithms give them the required information with an amazing level of accuracy and precision. The agents can also track the movement of the ship on the dashboard. Currently, there is an increased reliance of port agents and terminal operators on phone calls, emails, and notes. Dashboards help them in effective decision-making and they can also track if there are deviations from the optimum, desired performance. AI-powered technologies provide real-time information regarding the route recommended by the weather service, the ideal route, and the current route followed by the captain. Any deviations to the recommendations are also reflected on the dashboard.

Another key benefit of AI is in the domain of vetting. Ship owners and vessel operators need to have an assurance that the fleet will be accepted and acknowledged by the charterers. They do not focus on the quality improvements in the vessels to achieve the distinction. Instead, they focus only on those factors that are necessary for passing the criteria. The vetting process, however, is a lengthy work flow. The request routes at different authorities including port authorities, terminal authorities, and inspectors. Data analytics can automate all the process of vetting such that the relevant information is available to both ship owners, ship operators, and charterers. They can analyze the data extracted by the algorithms from different sources and select the most appropriate route that carries the minimal risk. In this context, the risk is assessed regarding the safety management, environmental safety, and navigation accuracy.

AI-powered systems also make it possible for ship owners, ship managers, maritime institutes and organizations to enable autonomy and remote monitoring in the system. Through the implementation of these systems, the business managers achieve remote diagnostics of the vessel operations. In the advanced systems, even remote piloting is possible and the whole system can be monitored based on a certain condition or criteria. Therefore, there can be only a limited presence of the crew on the vessel. The autonomy has also been achieved by introducing autonomous and unmanned vessels. Through the incorporation of these features, the risk to human life can be reduced and the operational cost can be decreased. As show in Figure 11 below, a large ship is in the operations in the sea. However, predictive diagnostics and preventive maintenance can be carried out at the base station. Moreover, routing and navigation may also be controlled remotely by sending the relevant data at the base station. Even the condition of engine can be monitored remotely through the use of sensors and the cloud data.

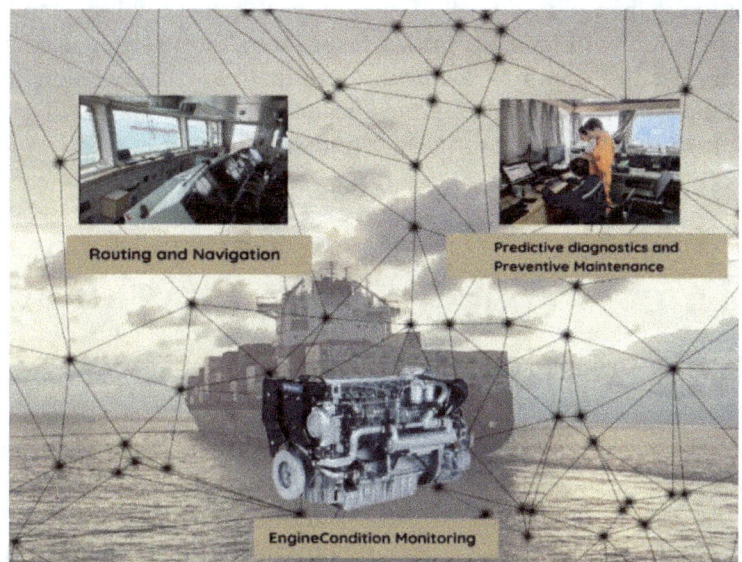

Figure 11: Remote Monitoring of the Vessel Operationsxxvi

1.2. CASE STUDIES OF AI-POWERED MARITIME TRANSPORTATION

The incorporation of AI tools into maritime shipping has already resulted in transformative changes in the maritime industry. There is an enhanced safety of maritime operations and all the operations in the supply chain have been optimized. In this section, I will provide you details of specific case studies so that you can get a sense of how real-world applications are being developed based on AI concepts in the sector of maritime shipping.

While reading these case studies, you should keep various aspects in mind. These AI models have been implemented by the renowned companies of the world. Therefore, the success stories provide a good way forward for the implementation in your maritime institutes and organizations. However, AI-based implementations will require extensive investment of material and human resources. Therefore, you should also consider the drawbacks of implementing AI technologies. My purpose is not to discourage you because I have written this book to advocate the use of AI in maritime shipping. However, as a responsible writer, it is my obligation to present to you both sides of the coin. There are both pros and cons of AI implementations and you should be well aware of both sides.

Figure 12 below summarizes the key benefits and drawbacks of using AI in the business world. The increased level of efficiency is no doubt the biggest benefit. In the case of maritime shipping operations, the efficiencies are achieved in fuel optimization, route optimization, and cargo handling optimization. The human risk is also reduced because many tasks are automated and executed by machine learning techniques. The decision-making process becomes intelligent and is backed by big data and historical data analyses. The availability of the safety mechanisms is also 24/7 because the machines replace the human responsibilities.

However, Figure 12 also highlights the drawbacks of using AI in the business world. The first drawback is that as soon as you implement AI tools and technologies by following the case studies that I have mentioned below, there will be a job insecurity and a resistance to change by the existing workforce. They will fear that machines will replace their valuable work and they will become redundant. Therefore, as I will explain in my next chapter on maritime business, the business managers will need to restructure the jobs before implementing AI in maritime shipping. They will also need to hire AI experts and make them permanent part of their workforce. The second drawback is the high initial investment. As the AI technology is comparatively new, the maritime shipping companies do not have the required equipment, server machines, cloud computing infrastructure, and sensors to implement AI systems. The implementation will benefit them in the long run, but initially the ship managers will have to convince the senior management that this investment is the need of the hour and they cannot do away with it. If there are no investments on the AI-based infrastructure, the business will no longer be sustainable. It's as simple as that!

Another drawback of AI implementation will be that there will be an increased reliance on machines. Although AI algorithms are highly powerful in predictive maintenance, but if the algorithms themselves fail due to a technical issue, then whole maritime shipping operations will come to a halt because the system will be totally reliant on AI-powered tools and technologies. Therefore, the ship owners, ship managers, maritime institutes and organizations should also develop risk mitigating strategies in this regard. Another issue in AI implementation is the reduced level of creativity. When everything can be handled and predicted by AI tools, the ship crew will have no work left for them. In the absence of AI tools, the ship owners and ship managers used to optimize routes through their own knowledge and experience. However, now, the value of their expertise will be limited because the crew staff will give more weightage to the results and recommendations made by AI algorithms.

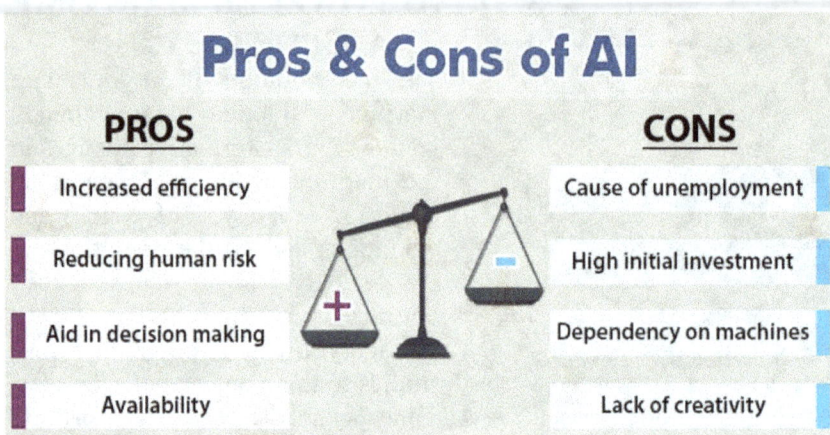

FIGURE 12: BENEFITS AND DRAWBACKS OF AI

The first case study is that of Rolls-Royce (RR). RR has an accomplished business at the international level related to marine systems and it is a prominent manufacturer of marine systems and marine engines. RR capitalized on AI-powered systems and introduced an Intelligent Awareness System on vessels. The system was based on a connected network in which sensors and cameras were used by the AI algorithms. The programs process and interpret the received data and facilitate the crew members by presenting to them a report of situational awareness. It is a highly beneficial report for the staff because it helps them in detecting other vessels in the sea, identifying the potential obstacles, and getting navigation recommendations for the voyage safety. RR reported that this AI-based intervention improved the performance of the crew significantly and they were able to make intelligent decisions for reducing accidents and collisions. As shown in Figure 13 below, the new AI-based awareness system is connected with the data center and transmitting all the essential data points to the center in real-time. The intelligence systems in this deployment are so powerful that they can even manage the operations of obstacle detection, navigation, and communications independently and autonomously.

FIGURE 13: AI-BASED SHIPPING INTELLIGENCE SYSTEMxxvii

Another important case study is the implementation of predictive maintenance by Maersk Line (ML). ML enjoys the leadership among the shipping companies and the company is also providing leadership in the deployment of AI-powered systems. The predictive maintenance is one area of maritime shipping where AI has been utilized successfully by ML. AI algorithms receive data from the sensors and perform a critical analysis of the current health of the ship. They predict the areas where the maintenance has become due. They also indicate the components and critical machinery that will be needed for carrying out the maintenance work.

The algorithms also forecast when the equipment will complete its useful life and result in a failed operation. It enables the ship managers to opt for scheduled maintenance. This strategy makes a significant reduction in the unplanned downtime, and there is an optimization of the overall operations of the ship. As indicated in Figure 14 below, through shipborne sensors, ML is collecting data from the vessels in real-time. This data will be used not only for predictive maintenance but also for route optimization, unloading the cargo at different ports, and gaining efficiencies while loading the cargo. As can be seen in Figure 14, the data that is being transmitted has a volume in 30 TB or 2 TB in some cases. It is a common scenario in AI-based and IoT-based systems. You can have a real value of your investment in the IT infrastructure if the sever machines at the data center are capable of receiving and processing big data. When you work in an IoT-based environment where the devices are 'talking' to each other, there is huge sending and receiving of data. The AI algorithms are well equipped of processing big data. However, they need high processing power machines for processing the data. Therefore, your IT infrastructure should be robust and capable enough for handling big data.

FIGURE 14: PREDICTIVE MAINTENANCE FOR VESSELSxxviii

The third case study is that of CMA CGM (CC). CC is a well-known company in the domain of container shipping. The company made use of robotic systems powered by AI technologies. This system is being used by the container shipping company for automated handling of the cargo. The robots at the company are equipped with AI systems. They have also been trained by using the computer vision technology. They perform the tasks of loading and unloading the containers from the ship without human intervention. It has enabled the company to reduce the staff cost and handle the cargo at the terminals efficiently. The real benefit that CC got from this intervention was a reduced time to handle the cargo that provided them with a competitive advantage. Moreover, there was a higher level of efficiency in the operations of cargo handling.

As mentioned in Figure 15, the loading and unloading of containers by CC is being performed by robots. It has been termed as a connected device solution by CC and CC launched this solution in collaboration with a French start-up known as Traxens. In this solution, the container also has a connected box. It enables the data center to know the position of the container, the temperature data, and the resilience level of the container to the potential shocks.

Figure 15: AI-Powered Container Loading/Unloadingxxix

The fourth successful case study is that of Wartsila (Wart). Wart is a provider of innovative technology solutions and the company has a proven track record in introducing technology-based marine solutions. The company developed a new product known as Smart Marine Ecosystem. This AI-powered system evaluates the data gathered from the vessel operations and uses it for the optimization of fuel consumption. The data is also used for recommending maintenance schedules and navigation routes. The data analysis also contributes to the operational efficiency and sustainability of marine ecosystem. There are various benefits of deploying Wart's system. The system will improve the carbon footprints record of the shipping company. There will be increased cost savings and the maritime operations will be more sustainable and efficient. Figure 16 below shows the marine ecosystem developed by Wart. As you can see, the system is connected to the satellite for connectivity and data transmission. The system also employs a good security mechanism to prevent the unauthorized access of data in the IoT environment.

Figure 16: Marine Ecosystem by Wartsilaxxx

Another important case study is that of Kongsberg (KB). KB offers advanced IT solutions for the maritime shipping industry. KB developed a DP (Dynamic Positioning) system for the ships that was powered by AI technologies. The system capitalized on the data received from multiple sensors configured on the ship. The data processing by AI algorithms made it possible for the ship operators to maintain an accurate position of the vessel. It was particularly useful during challenging weather conditions when it was required to maintain the stability of the vessel. The risks in the critical vessel operations reduced significantly by this deployment and the operational efficiency was enhanced remarkably. As highlighted in Figure 17 below, the system developed by KB considers various parameters affecting the vessel operations. These include wind, waves, sway, tunnel thruster, and others. Based on the evaluation of all these parameters, the accurate positioning of the ship is recommended to the captains.

Figure 17: Dynamic Positioning Systemxxxi

Different companies have also designed AI-powered systems that can transform the future of maritime shipping. Here I am mentioning some of these solutions so that you can realize how much efforts have already been put by the technology and electronic companies. There is a need for the ship owners, ship managers, maritime institutes and organizations to realize the potential of these tools and gadgets. They should incorporate these tools in their organizational operations to gain a competitive advantage.

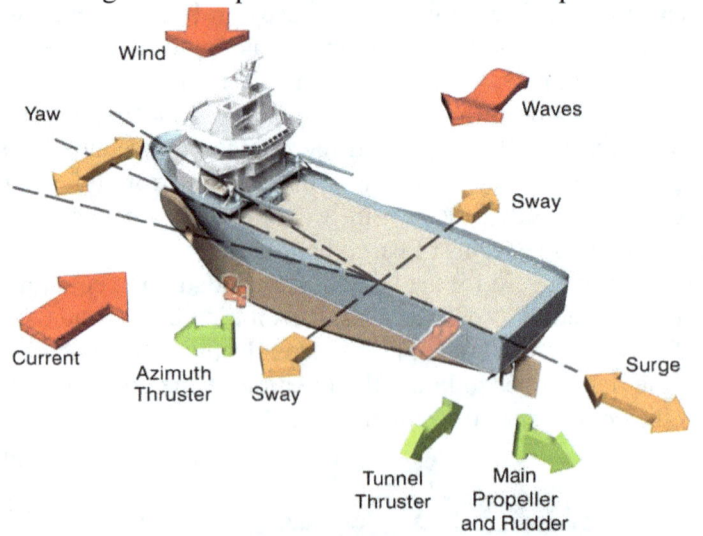

Hitachi is offering an innovative solution for the maritime organizations where the AI-powered system can produce impressive results in the optimization of fuel consumption. AI algorithms evaluate the historical data of the vessel. The data points analyzed by the system include sea state, weather patterns, and engine performance. Based on these parameters, the system recommends the appropriate speed and route for the ship. As shown in Figure 18 below, physical space is just one layer in this implementation that is made of vessel information, camera information, and cargo information. The next layer is the data layer where on-site work is carried out by using big data and sensor data. The software provided by Hitachi also optimizes the storage plan of the container and checks the container damages.

FIGURE 18: HITACHI'S SYSTEM FOR MARITIME TRANSPORTATIONxxxii

Mitsui O.S.K. Lines (MOL) is a Japanese company offering AI-based maritime solutions. The company offers a route optimization system for the maritime industry. The system achieves route efficiencies by training the AI algorithms from the historical data. The optimized routes are recommended based on the factors of fuel prices, traffic congestion, and weather patterns. The outcomes are achieved by the shipping companies in the form of reduced fuel costs and reduced voyage time. As shown in Figure 19 below, the vessel allocation by MOL gets data from multiple sources. The vessel allocation is recommended by simulating transport routes per vessel.

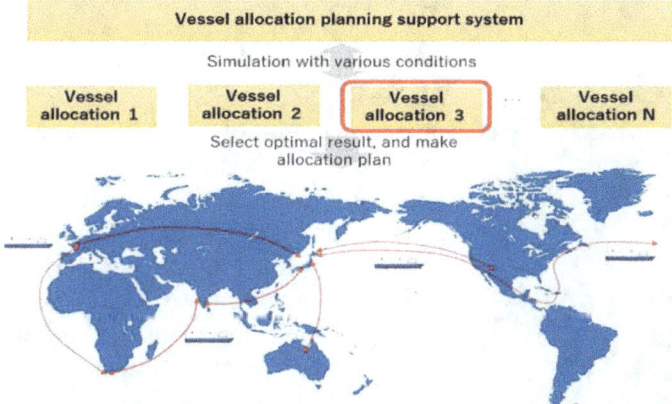

Figure 19: Vessel Allocation Plan by MOLxxxiii

Another prominent solution for the maritime industry is offered by Awake.Ai. Awake is a Finnish organization. The company offers an AI-based time data system. This system processes historical data by using machine learning techniques. The vessel data is used to identify the patterns in the data and predicting the future state of the vessel. The information provided by the time data system is used by the ship managers for routing and scheduling. As highlighted in Figure 20 below, the key features of the system are time of arrival prediction, destination prediction, resource tracking, and cargo operations prediction. In the cargo operations prediction, the ship data, cargo data, and pilotage data are analyzed. The evaluation uses anchorage model, congestion analysis, regression analysis, and data annotation.

FIGURE 20: TIME DATA SYSTEM BY AWAKE.AIxxxiv

A closer look at the model of awake.ai presents further interesting details. At the bottom of the model, the system parameters are mentioned. As is evident from these parameters, AI-based systems do not rely on a single source of information but obtain data from multiple sources to improve the recommendations. Therefore, the recommendations in this case are based on a combination of AIS data, port call data, ship data, cargo data, weather data, pilotage data, and data from cameras. The port call data is used for historical analysis and string classification. AIS data is also used for trajectory prediction. The weather data is used in sea voyage model and pilotage model. The ship and cargo data further expands the analysis to anchorage model. The pilotage data is used for congestion analysis because the collisions usually occur in congested areas. Therefore, the congestion analysis is crucial in vessel voyage. The system also develops regression models based on the data of cargo and operations so that the contribution of each variable could be ascertained in predicting the successful cargo operations. The data from the camera devices is used in the computer vision systems of AI. As shown in the top layer of Figure 20, all these analytical systems produce four types of prediction. The first prediction is regarding the destination as to when the cargo is expected to reach the destination considering the weather patterns and the climatic conditions. The second prediction is regarding the exact time of arrival and this prediction is based on trajectory, sea voyage, pilotage, and anchorage models. The third prediction is regarding the cargo operations. The cargo should reach the destination timely and in good condition. The fourth prediction is regarding the tracking of the resources in the cargo that is accomplished based on regression models and computer vision.

IBM is also one of the technology leaders in offering solutions to the maritime shipping. IBM has recently announced a project that is still under the development stage. However, the maritime shipping professionals are already excited about this development and they want to use the features offered in this new and innovative project. This project has been named as Mayflower. The Mayflower ship will be supported with various AI sensors. The ship will navigate the sea autonomously without human intervention. This project has the potential of transforming the future of maritime shipping. As indicated in Figure 21 below, the Mayflower 400 ship is unmanned, and it is navigating the sea autonomously.

FIGURE 21: UNMANNED MAYFLOWER 400 BY IBMxxxv

1.3. USING AI IN VESSELS

I mentioned a lot of successful cases in the previous section. From these successful models, you will have realized that AI has the potential of transforming the future of maritime shipping. When you use AI in vessels, you will have to consider several factors for a successful implementation. By now, from the basic concept of AI, and the examples from the real-world, you are well aware that AI algorithms are as good as the training data available to them. For example, if the robots are not trained for the loading and unloading of cargo, they will not be able to differentiate between containers and other materials.

Computer vision is an integral component when AI is used in vessels. The robots should know which instructions are being given by humans that they have to follow and which machines should be used for executing the tasks. It is not just two or three instances in the computer vision that are sufficient for a robot to train. The robots are given thousands of true and false examples so that they know the good behaviors, bad behaviors, and the expected behavior. Figure 22 below shows some of the examples in Computer Vision that a robot will need to make a sense of the maritime environment.

FIGURE 22: OBJECT DETECTION BY ROBOTSxxxvi

Another challenge in the vessel operations is to remain connected with the data center. As I explained through the case studies, the AI-based systems are controlled by a remote center that processes the data received from the sensors configured on the vessel. The whole marine ecosystem works in an IoT environment in which the devices are connected to one another. But who connects them? Well, the answer is pretty straightforward. The connecting medium is internet. However, the next question is: Can I use a wired connection in the sea. In most of the cases, it will not be possible for you to use a wired, high-bandwidth internet connection. As shown in Figure 16, the marine ecosystem by Wartsila uses a satellite-based internet connection that is the most viable option for a good speed and sending and receiving data at a higher rate. The satellite-based internet connections are quite expensive these days. Therefore, the ship managers will have to consider all the cost savings that I have mentioned in the above sections. AI-powered systems will reduce fuel consumption, optimize the voyage routes, and reduce the downtime of maritime operations. When the proposal of AI implementation is presented to the senior management, these factors should be mentioned in the proposal so that the management is convinced with the utility of the proposed systems. In the absence of a cost benefit analysis, the management might feel that the cost of IT infrastructure is increasing because the current IT infrastructure of the shipping company will need a complete overhaul for a successful AI transformation.

Another area of consideration is the quality of human resources in the shipping company. AI experts and AI-offering organizations will no doubt deploy an impressive system in your organization. However, after the deployment, these systems are to be run and executed by the in-house staff. Therefore, the training and development of the in-house team is highly crucial for a successful implementation. AI algorithms are highly complex and only computer scientists and programming experts can decipher the working of these algorithms. As an end-user, the maritime shipping professionals should focus on the outputs, reports, and dashboard indicators that are facilitating them in maritime shipping operations. These outputs should be used intelligently for transforming the future of maritime shipping.

The shipping companies will also have to hire a few IT professionals as permanent employees. It is because the issues may arise in the deployed AI systems and an IT team will be needed for maintaining the systems, server machines, and the repositories. In an ideal case, some organizations develop systems from the in-house IT team. However, it might prove highly expensive because as I explained, the AI-based systems for the maritime industry are based on sophisticated technologies that will be highly challenging for the in-house IT team. Moreover, these solutions are not tested and they might deviate from the industry norms and standards. Therefore, a better solution is to procure the state-of-the-art solution offered by different vendors. In the earlier section, I have mentioned some of these vendors such as Hitachi, Mitsui OSK, Awake, and IBM.

1.4.   AI FOR PREDICTIVE MAINTENANCE

When predictive maintenance is discussed in the context of AI, it should be noted that it is an evolutionary concept and its significance was felt when the previous maintenance strategies could not give the desired results.

FIGURE 23: EVOLUTION OF MAINTENANCE STRATEGIESxxxvii

As shown in Figure 23 above, in the first era, the reliance of the industrial and commercial organizations was on reactive maintenance. As the saying goes, 'if it is not broken, don't fix it.' So, in this strategy, the technicians intervene only when there is a breakdown in the system and the equipment fails. The failure might occur during the off-timings of a factory or it may also occur during the regular operations of the business. The reactive mode was not considered as a prudent strategy because the customers expect the timely delivery of their orders and they are not concerned about the failure of the equipment or the breakdown of the system.

The next approach that was a comparatively better strategy was the preventive maintenance. In this strategy, the machines and equipment went through a scheduled and regular maintenance. Due to the periodic servicing of the parts and the overall system, there was a lesser likelihood that a machine will fail at the critical time of the operations. However, the failures still occurred during the time that elapsed between one scheduled maintenance and the other scheduled maintenance.

The third strategy that was powered by the AI-based systems is popularly known as predictive maintenance. In this strategy, the equipment and machines are connected with sensors. The sensors continuously transmit the data regarding the current state of the machine to the software applications. The software then, based on the learning and the receiving of the current data, predicts the possibility of the equipment failure in the future. It is a highly beneficial information because let's say the result is that one part of the vessel will fail in the next one month. Then, the ship managers can start the maintenance work or the replacement work of that part from now. On the other hand, if the ship managers ignore the alerts of the software, they will be held responsible for the losses suffered by the organization. It is because the reports of the software applications are also accessible to the senior management. These reports will be used to identify the responsibility if any incident or unfavorable event occurs.

Figure 23 highlights that the world has progressed even one step ahead of predictive maintenance. The AI-based algorithms are now capable of performing prescriptive maintenance. In this strategy, the machine learning techniques are used not only for predicting the future failures of the equipment or machine but the algorithms also recommend solutions for avoiding or rectifying the failures. It is just a simulation of the doctor's room, where the patient is not only diagnosed for the issues in the human body but the doctor also prescribes the medicines for recovering from the illness and maintaining the health and wellbeing.

As highlighted in Figure 24, predictive maintenance is a function of time and shipping condition in the context of maritime shipping. The time points vary based on several factors. It depends on whether the predictive maintenance is being carried out by using monitored data. It might also be influenced by the frequency of planned maintenance. The approach is considered highly reactive when the shipping companies repair the equipment only when it is about to fail.

The shipping conditions are reported when the fouling in the vessel system grows. It is also flagged when the engine output is reported at a reduced level. Another criterion is to flag an alert when the highest speed of the vessel is never achieved. A red flag is generated when there is a catastrophic failure or the vessel system appears to be in adverse condition.

FIGURE 24: PREDICTIVE MAINTENANCE IN MARITIME SHIPPINGxxxviii

1.5. AI for Cargo Optimization

In the maritime shipping, the main area of business is to transport the cargo from one destination to another destination through the sea route. Therefore, the senior management of the shipping companies will welcome any solution that optimizes the performance of cargo delivery.

To understand the role of AI in cargo optimization, it is important to understand the whole shipping process. As shown in Figure 25 below, the shipping process begins with the shipper and ends with the consignee. Between these two extremes, there are various touchpoints including origin warehouse, origin port, destination port, and destination warehouse. The vessel operations are limited to origin port and destination port. The transportation to other touchpoints is through trucks or other road vehicles. AI-based systems not only automate the process at all these touchpoints but also optimize the route for a safe and quick delivery.

FIGURE 25: THE PROCESS OF SHIPPINGxxxix

As indicated in Figure 15, the loading and unloading of cargo at different touchpoints can be accomplished by using robots. Moreover, the sensors attached to the containers can also reveal the quality of the products in the container. Based on the weather patterns and climatic conditions, if there is a possibility of damage to a product in the container, the alerts will be issued promptly by the AI-based systems. The autonomous ships such as the Mayflower by IBM (Figure 21) can help in shipping food and other accessories to the areas where the human lives are at risk due to war-like situations such as Ukraine, Yemen, or Afghanistan.

FIGURE 26: AI FOR CARGO OPTIMIZATIONxl

Figure 26 above shows that when AI is used for cargo optimization in maritime shipping, various benefits are reaped by the organizations. AI-powered systems predict the demand with accuracy and precision. Therefore, the vessels can be allocated and reserved accordingly for maritime shipping. The algorithms can also predict the peak hours and higher sales days during which the company may reduce the cargo charges and benefit from the economies of scale. The best route for the vessel operations can also be found. The customers can also be updated regarding the receipt of the goods with a higher level of accuracy. The vehicle routing outside the port operations is also efficient and ensures the last mile optimization. With the help of AI-based systems, the ship managers can be confident that the cargo will reach the customers safely and there will be more and more happy customers.

1.6. AI FOR NAVIGATION AND ENHANCED SAFETY

The navigation tools are also crucial for the maritime shipping because the vessels do not travel in the sea in isolation. Other vessels also operate in the same environment. The goal of the ship captain is to ensure a safe voyage of the vessel and deliver the cargo within the committed time. When AI-based systems are used, several factors should be considered for optimizing the navigation and safety of the vessel as shown in Figure 27 below. The consideration of these factors together creates a FOCUS (Fleet Optimal Control Unified System). As indicated in Figure 27, the sensors capture various data points on the vessel. These include the cargo, hull, and the condition of the engine. The data points also include sea conditions and the weather information. The information regarding the other vehicles is also sensed by the algorithms. The satellite connection is also capable of receiving voice data and visual images. The ship's log maintenance record is also available to the data center.

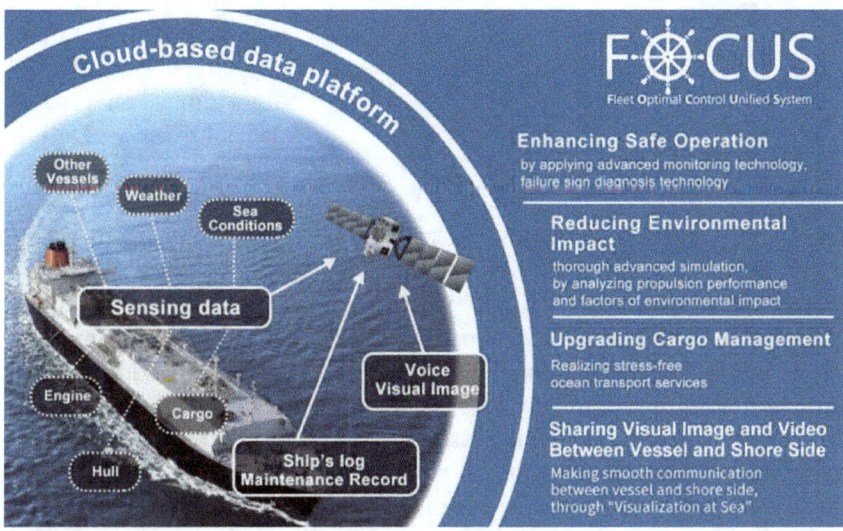

FIGURE 27: FOCUS SYSTEMxli

When the cloud-based data platform produces a FOCUS system, four key benefits in vessel operations are achieved. The vessel operations become highly secure because advanced monitoring technology is used in the system. The predictive maintenance system flags alerts regarding any failure signs. Another benefit is that the environmental impact of the vessel operations is reduced. The AI algorithms evaluate propulsion performance, evaluate the impact of the vessel operations, and prescribe the best route to reduce the environmental impact of the operations.

Another benefit is an efficient management of the cargo. The ocean transport services are delivered efficiently and the status of the containers is also reported at regular intervals to the data center. A good aspect of the FOCUS system is that the AI systems are also able to process voice and image data. It facilitates a smooth communication with the crew staff and the sea conditions can be visualized effectively at the shore side. The FOCUS project was introduced in 2019 and it was well received by MOL. MOL installed fleet viewer application in May 2019xlii. Through this project, MOL was successful in collecting almost 6000 items from the vessel data. The data was transmitted to the shore center at a very high frequency of one minute intervals.

Figure 28 shows that different factors affect the performance of AI-based navigation systems. An ideal system should have a touch screen so that the user can have a reality experience of the voyage. The vision should be available with both thermal camera and day camera. The dashboard indicators should be integrated with the existing sensors. The computing power of the system should be very high. The systems should have the capability of updating and improving them through artificial intelligence and deep learning.

FIGURE 28: FACTORS AFFECTING THE ACCURACYxliii

1.7. AI FOR FUEL EFFICIENCY AND OPTIMIZED ROUTES

Fuel efficiency has become a critical factor in airline operations as well as maritime operations. It is because the energy cost is on a constant rise. If the fuel is not used efficiently, the cost of maritime shipping will become extremely higher and the business will no longer be sustainable. Fuel efficiency is also crucial due to the fact that environmental protection agencies blame maritime organizations to be significant contributors in carbon emissions. The emissions can be reduced by the efficient utilization of fuel. The optimized routes are equally beneficial in maritime operations because if the ship reaches the destination by travelling a reduced distance, then fuel will be saved.

Maritime operations are highly complex and it can be life-threatening if fuel saving and route optimization strategies are left to the discretion of ship captains. There should be a use of scientific

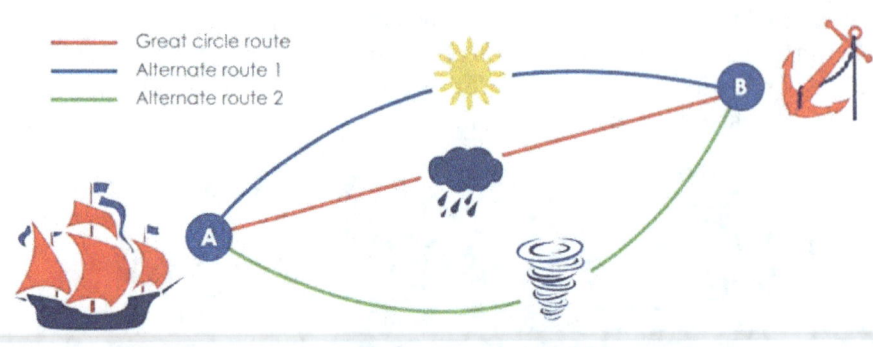

techniques for enabling fuel efficiency and route optimization in maritime shipping operations. AI is definitely the best option for accomplishing these objectives because AI algorithms can process data from multiple touchpoints and assist in intelligent decision making. In the below paragraphs, I will explain to you how AI can be a game changer in maritime transportation concerning route optimization and fuel efficiency.

When it comes to fuel efficiency, AI algorithms evaluate various parameters in the maritime environment to recommend a solution for an optimized fuel consumption. Historical fuel consumption reports are used by the machine learning techniques as the training dataset. The algorithms even recommend the optimal configurations for the vessel engines so that the fuel efficiency could be accomplished without compromising the performance of the maritime operations.

When it comes to route optimization, AI algorithms rely on the data of sea conditions, weather parameters, fuel cost, and traffic parameters. The details regarding these variables are processed for recommending optimized routes for the vessel. It should be noted that AI implementations may take the form of recommendations or the process may also be automated. In the case of recommendations, the final decision is still made by the ship captain. However, in the case of automation, the recommendations are automatically implemented by the AI-powered systems in real-time. When AI-powered systems are used, the routes adopted by the vessels are the most efficient and the system is in a position to avoid adverse weather conditions.

There are also other advanced implementations of AI concepts. The algorithms may also process data received from the meteorological department and make weather forecasts in an accurate manner. It facilitates the captain to modify and improve the planned route. The algorithms can also help in the planning process of maritime voyage. By considering the variables such as vessel specifications, fuel cost, environmental laws, and cargo load, a better voyage plan can be developed by the algorithms. The fuel consumption can also be tracked in real-time by integrating IoT sensors to the AI systems. In this way, AI analytics presents a holistic view of fuel consumption and route situation.

From the above key implementations, it becomes quite evident that AI is the way to go for implementing fuel efficiency and route optimization in maritime shipping. The ship owners, ship managers, maritime institutes and organizations may expect that there will be further improvements in AI-based systems for fuel efficiency and route optimization. It will eventually contribute to an efficient and sustainable future for the maritime shipping. Figure 29 below shows an AI implementation where the AI-powered systems have proposed three routes for the vessel voyage. The planned route is the great circle route that has been marked with red color. It was planned by the ship managers because it shows the straight line and the shortest distance between the origin A and the destination B. However, the weather conditions have made this route unfavorable. There are risks of a high storm and pirates due to the influence of environmental variables. Therefore, the algorithms have suggested two other routes. Route 2 is marked with blue color and it shows a good sunlight from the weather data. The voyage can easily bypass the weather conditions but it will lead to consuming higher fuel and arriving late at the destination B. Route 3 is marked with green color. Although in this route, there are no rainy conditions, but there will be some episodes of drift. The weather will be comparatively good and the fuel consumption will be less than Route 2. This option may also be adopted by the captain. It involves a delay in reaching the destination B, but there is a significantly less consumption of bunker fuel. These comparisons by AI-based algorithms are highly beneficial for ship owners, ship managers, maritime institutes and organizations.

FIGURE 29: AI ALGORITHMS PRESENTING 3 OPTIONSxliv

Another successful case study of route optimization and fuel efficiency is that of Yara Marine (YM). YM implemented Route Pilot AI in September 2022, which is based on the concept of digital twin. It is a cloud-based AI system in which high computing power is utilized for processing high frequency data. Based on these calculations, the most optimized propulsion settings are developed for the future voyages. As shown in Figure 30 below, the digital twin model simulates all the parameters of the voyage operation. It enables the shore-based staff and the ship crew to arrive at the optimal working parameters. The net result is an energy-efficiency journey and a reduced consumption of fuel. The algorithms rely on historical vessel data and sea parameters for arriving at the peak and optimized operational parameters. The system also improves the communication between the ship crew and the shore teams.

FIGURE 30: FUEL EFFICIENCY BASED ON DIGITAL TWINxlv

Figure 31 below shows an AI implementation where route optimization has been accomplished for maritime shipping. The system works on the premise that the straight line route is not always the ideal route and the ship managers should also consider the impact of other internal and external variables. An ideal vessel route for an AI-based system is the one that is saved from storms, heavy downpours, underwater currents, and high waves such that the performance of the crew is not affected and the possession of cargo is secure. The route should also reduce the carbon footprints of maritime institutes and organizations. The implementation in Figure 31 is based on AI algorithm developed by Sinay. The system works based on the modern route planning features. The system is tightly integrated to a centralized server that provides real-time data to all stakeholders. The meteorological changes are also taken into consideration. The system also processes data from different sensors for facilitating in the vessel route planning. This AI implementation is promised to provide fuel efficiency as well as the vessel safety. The forecasting model predicts the future state of the vessel with a high level of accuracy and precision.

Figure 31: AI Implementation for Route Optimization by Sinayxlvi

From the above implementations, it is evident that AI-powered systems have already been implemented successfully for fuel efficiency and route optimization in maritime shipping. These companies are going to gain a competitive advantage and lead the industry in the future. Therefore, other ship owners and ship managers should also evaluate these AI-powered systems with a long-term orientation. In the short-run, the implementation of these systems might appear costly. However, when these systems are implemented by every other maritime shipping company, then these AI-powered systems will become a necessity for the maritime institutes and organizations. It is high time that the power of AI is unleashed by ship owners, ship managers, maritime institutes and organizations. The crew staff and shore team will also need an extensive training for utilizing the new systems. It is because as I wrote earlier, AI-based systems can be in the form of recommendations or they can also be automated. In the earlier implementations, you are most likely to use the recommendations of AI-powered systems. Therefore, the end users should be well trained to make the optimum use of these recommendations that are based on big data analysis and real-time assessment of all weather-related parameters.

1.8. AI FOR PREVENTING ACCIDENTS AND COLLISIONS

Avoidance of accidents and collisions has become extremely critical in maritime shipping. The vessel voyages are highly expensive journeys and no maritime institute or organization can afford to bear this loss. The situation is particularly critical in the context when several vessels were found to be involved in human trafficking. There was a massive load of passengers on this vessel who were going to countries of opportunities for better living conditions. However, these vessel operations were not sustainable and citizens of many countries lost their lives in these vessel journeys. For a standard and a highly reputed maritime organization, it is crucial to implement every possible measure to prevent accidents and collisions because these accidents will immediately be known in the public domain and will have a detrimental effect on the reputation and goodwill of the organization.

AI-powered systems are playing a significant role in mitigating the risks of collisions and accidents. I am mentioning here several disruptive systems and technologies that can prevent collisions in maritime shipping.

The collision avoidance systems have been introduced based on AI technologies. These systems collect data from various gadgets connected in the maritime ecosystem. These may include surveillance cameras, sensors, radar, and identification systems. The aim of deploying these gadgets is not only to monitor the working of the vessel but also to evaluate the surrounding environment of the vessel. The machine learning techniques also present alerts regarding various collision risks that emerge during the vessel journey. The algorithms may also be implemented in a way that there is an automatic course adjustment and there is a minimal risk of collision.

Another state-of-the-art technology is situational awareness and situational monitoring of the vessel operations. AI algorithms evaluate different parameters to optimize the situational awareness for ship owners and ship managers. These parameters include the traffic density, the current position of the vessel, navigational hazards in the surroundings, and the current weather conditions. Based on the recommendations of AI algorithms, the ship managers can navigate safely and make informed selection of choices.

Collisions and accidents may also occur due to the deteriorating condition of the vessel. Therefore, predictive analytics and anomaly detection tools of AI also play a significant role in avoiding collisions. AI-powered systems can highlight the irregularities in the behavior of vessel. When the algorithms learn the standard safety protocols, they can indicate deviations from these protocols even if a single part or equipment deviates from these protocols. It enables the ship owners and ship managers to introduce proactive interventions and avoid accidents and collisions.

AI-powered systems also work exceptionally well in adaptive learning. The algorithms are involved in the process of continuous learning and improve their knowledge based on past accidents and near-miss collisions. The risk assessment model then becomes highly effective and prediction accuracy of the AI-powered system is improved.

As indicated in FIGURE 32 below, AI-based collision avoidance system is providing a good situational awareness to the captain. According to a study, the biggest reason of vessel collisions is that water becomes congested with a large number of vessels coming closer. It is particularly experienced in several key places of the vessel voyage such as when the vessels are in coastal waters, approaching port, in the anchorage, or are present in traffic separation areas. Therefore, situational awareness is a key to avoiding collisions in congested areas. The ship captain should also be aware of what is happening in the surrounding of the vessel and what the consequences will be of his inaction.

FIGURE 32: SITUATIONAL AWARENESS FOR REDUCING COLLISION RISK

The implementation mentioned in FIGURE 32 uses deep learning algorithms. Marine targets are detected by these algorithms in real-time. There is a classification and tracking of the risks that provides an overall situational awareness to the ship captain in real-time.[xlvii]

AI-based safety systems have not only been designed for manned vessels but also for autonomous ships. One of the leading cases in this regard is Orca AI. The system developed by Orca offers improved visibility of the vessel in critical conditions as shown in Figure 33 below. The system is also capable of detecting those vessels that cannot be seen by human eye at night. Thermal cameras and vision sensors have been used for this purpose. The algorithms analyze the sea environment and generate alerts regarding the dangerous situations. The system developed by Orca can also detect those small markers and fishing boats that a radar fails to capture.

FIGURE 33: COLLISION AVOIDANCE SYSTEM BY ORCA AI[xlviii]

Another good implementation of vessel collision has been accomplished by Fujitsu. The system developed by Fujitsu is based on the prediction technology of AI. The efficiency of the system is demonstrated in Figure 34 below. The control operations are significantly optimized by the use of Fujitsu technology. In the first evaluation, the conventional method has been depicted. In this method, the hazards and risks were determined by the ship managers based on their own knowledge, skills, and experience. However, in evaluation two, the conventional methods were used in conjunction with AI-based risk prediction technology. The algorithms made use of risk recommendation logs and historical AIS data. It was found that the use of AI technology significantly reduced the time before a risk warning is issued. The number of warnings also increased twice by using Fujitsu AI-based system. The technology proved particularly significant for newcomers in the ship crew. They implemented control actions for risk avoidance with the same level of effectiveness as the highly-skilled controllers. Therefore, the AI technology also assisted in bridging the skills gap between newcomers and experienced controllers.

FIGURE 34: COLLISION AVOIDANCE BY FUJITSU TECHNOLOGY[xlix]

As I demonstrated through the above three examples of AI in collision avoidance and accident prevention, the use of AI is highly beneficial in this domain. Maritime shipping incurs double losses when there are collisions and accidents. The ship owners, ship managers, maritime institutes and organizations not only bear losses of their valuable property but they also have to pay for the penalties and damages inflicted to the cargo. Some customers may also opt for court cases that further increase the cost of damages to the maritime organizations. Therefore, maritime institutes and organizations should consider implementing AI for preventing accidents and collisions.

This discussion concludes our journey of chapter two. In this chapter, I presented to you theoretical concepts and real-life examples of how AI can be beneficial in maritime shipping. You will have realized by now that AI has an immense potential for transforming the future of maritime shipping. I highlighted to you various successful cases in section 2.2. Then I explained to you specifically for every vessel operation one by one, how AI is being used successfully for optimizing those operations. AI is transforming vessel operations and is the most appropriate tool for predictive maintenance. AI is great for cargo optimization. It is an ideal tool for navigation and voyage safety. I also demonstrated how the implementation of AI can give fuel efficiency and route optimization to ship owners, ship managers, maritime institutes and organizations. Lastly, I focused on the role of AI in avoiding accidents and collisions. Now you know about AI in maritime shipping in and out. It is an ideal time to move on to the next chapter where I will tell you how you can use AI in maritime businesses, business processes, and business operations.

2. AI AND MARITIME BUSINESS

You might feel that I have already deliberated on the implementation of AI in the maritime business. But my answer will be No. In Chapter 2, my focus was on optimizing the operational side of maritime shipping through AI. Therefore, I demonstrated how AI can be used in vessel operations, predictive maintenance, cargo optimization, navigation, safety, fuel efficiency, route optimization, and collision avoidance. This chapter provides you with another perspective, and i.e. the business side of the maritime business. The chapter will highlight different managerial dimensions of AI implementation and what areas of decision-making can be influenced and optimized by AI.

AI is blamed for its disruptive nature and one of the biggest critic on AI is that it is reducing the employment opportunities in the business world. This allegation is also heard in the maritime shipping industry. However, when AI is shrinking the conventional job market, the technology is also opening new jobs that can easily be acquired by learning the AI tools and technologies. I have used the word 'easy' because AI is a comparatively new technology and if you even get a fair level of knowledge of different AI-based jobs, you are most likely to be hired by an esteemed organization. The ship owners, ship managers, maritime institutes and organizations should also expand their current workforce to hire AI experts in their teams. It is because when they implement AI-based systems, AI developers will come once to install the setup. Afterwards, they will only make periodic visits. A proper maintenance and utilization of AI-based systems will also require that you have a good skills base in your workforce as well. I have mentioned below some of the job titles that will be in extremely high demand when AI-based systems will be implemented extensively in maritime shipping industry. The ship owners, ship managers, maritime institutes and organizations, while hiring the new workforce, should pay particular attention to these job titles.

The first job holder you will desperately need in your maritime shipping business is an AI researcher. It is because the AI technology is continuously evolving. As I mentioned in the previous chapter, AI-based systems are being developed at three levels known as ANI, AGI, and ASI. There is a continuous evolution of the AI technology. Therefore, it is essential to hire the services of AI researchers who will keep the maritime institutes and organizations informed regarding the latest developments in the field of AI in maritime shipping.

The second job holder maritime institutes and organizations should be looking for is the expert in natural language processing. As I explained to you in the previous chapters, AI systems that only respond to stimuli are called reactive machines. This concept was popular in the initial days of AI but now it has become obsolete. Now those AI systems are in demand that work on the principle of continuous learning. The algorithms learn from the big datasets and these datasets may also contain textual data that can only be parsed by the experts in natural language processing. Think of the user generated content on social networking sites as a valuable data for a particular brand. The brand managers will like to process this data and generate summarized reports to know the differentiated aspects of their products and the areas of improvements. The statistical tools such as SPSS and Excel can only process numerical data. Natural language processing is a complete science in itself in which tokens are generated for parsing the data and generating the summarized reports. The maritime institutes and organizations will need these experts in their organizations to gain insights from the big data.

The third critical job holder is a prompt engineer. As the saying goes, 'we are living in a world where you are not required to give answers. All that is required is to ask good and relevant questions.' This proposition applies completely to the AI world where asking the relevant questions and requesting the appropriate data has become a scientific approach and is termed as a prompt. Its simplest examples are prompts for ChatGPT and Google Bard. Many organizations and technology experts are publishing a summary of good prompts and these prompts are read with a keen interest. It is because people have now realized the significance of good prompts. A prompt engineer is hired for designing a prompt. These prompts or requests can then be sent to the AI tools that the organization has acquired. The organization will be able to extract valuable insights from the AI systems when the prompt engineer will send effective instructions to an AI tool.

The fourth job holder is an expert of robotic process automation (RPA). When maritime institutes and organizations implement robotics in maritime shipping such as cargo optimization, this job holder will be required for RPA. The individual will assist the organization in different tasks such as data transfer, updating of customer profiles, data entry, inventory management, and other process automations.

The fifth crucial job holder is an algorithm auditor. It should be noted that although AI systems are automated and the learning process also takes place in an automated manner, but the developers of AI algorithms are humans. Therefore, human biases may affect the functionality of AI algorithms. For example, if an anomaly is to be detected in a machine, and the AI developer builds an alliance with the machine developer and does not detect the anomaly in the AI algorithm, the anomaly will go undetected due to human bias. Therefore, it is critical to hire an algorithm auditor who will go through all the details of the AI algorithms and ensure that they are free from biases based on color, race, or age.

The sixth and the last crucial job holder is expert in AI laws and ethics. Whenever any new AI technology emerges, ethical concerns are also raised regarding its usage. For example, when chatbots were introduced by Open AI, the academic community criticized the initiative that it will spoil the creativity of the students and lazy students will use it as a tool for creating their assignments. The turnitin software also included the functionality of AI detection besides checking plagiarism. In the maritime sector too, AI based navigations are being criticized because the navigation tools may also enter the sensitive and prohibited areas. Therefore, when maritime institutes and organizations implement AI in maritime shipping, the expert in AI laws and ethics will tell them the ethical implications of the proposed implementation, and then it is up to the management to go for the implementation after analyzing all pros and cons.

I have discussed these job roles at the very beginning of my maritime business chapter so that you could appreciate and realize how the entire maritime shipping business will be transformed by AI. Every business perspective will be influenced by AI implementation. It is now up to the senior management to make investments on the optimization of human capital as well so that the true benefits of AI implementation could be reaped by the organization. If the focus of the business managers is only on setting up the IT infrastructure and the significance of human resources is overlooked, then the real benefits of AI implementation will not be achieved in maritime shipping.

2.1. BENEFITS OF AI IN IMPROVING MARITIME BUSINESS

Maritime businesses will be benefitting from the AI in the long run. It is because AI-based systems are improving their efficiency and precision ratio and it will take some time when these systems will be available with their full potential. Moreover, as I mentioned in the introduction, the maritime institutes and organizations lack the required human resources for the maintenance and continuous usage of AI-based systems. These new job holders (particularly six that I mentioned) should be hired by the maritime organizations for the smooth running of AI-based systems. When there is a good time to implementation such as at least one year, the ship owners and ship managers will see improvements in the maritime business.

As shown in Figure 35 below, the improvements will be observed in the maritime business for shipping carriers as well as freight forwarders. There are five key areas where the maritime institutes and organizations will benefit from the AI implementation. The first area is the safety of the vessel voyage. AI will improve the safety of the vessel voyage by incorporating predictive maintenance and responding to the risk factors in the automated manner. The second area is the optimization of the vessel's capacity so that the maximum cargo could be loaded that can be sustained by a ship. For this purpose, AI-based systems will be supported with fuelled positioning systems and computer vision systems. The third area where the maritime businesses will flourish through AI is the introduction of dynamic pricing for cargo and freight. The route forecasting algorithms will tell the businesses how much they can lower the price while still maintaining the profits through the economies of scale. The fourth powerful outcome will be the route optimization and consuming the minimum possible fuel. It will be a huge savings for the maritime institutes and organizations because fuel is a major cost these days and puts a heavy load on maritime shipping operations. The fifth benefit is the enhancement in the productivity of the business processes whereby the AI tools are used as scheduling tools and advanced planning tools.

FIGURE 35: BENEFITS OF AI FOR THE BUSINESSl

You should keep in mind that AI is just a conceptual framework. It heavily depends on the skills and expertise of the AI developer how he utilizes the AI concepts to develop AI-powered system. The use cases of AI are innumerable and AI will advance technology and innovation in the domain of maritime shipping. The predictive capabilities of the maritime shipping will improve significantly and the vessel operations will be more efficient and effective. The ship owners, ship managers, maritime institutes and organizations should capitalize on the opportunities offered by AI-powered scheduling systems, real-time analytics systems, and automated systems for weather forecasting.

When AI-powered advanced analytics systems are used, the valuable business insights are gained in the maritime shipping because the data is processed from multiple sources. When the decisions are based on data-proven methods, the risk factors are reduced. Moreover, the ship owners and ship managers can provide a more valid justification to the senior management as to why they preferred a particular solution over other solutions. Another benefit of AI in the maritime shipping is in the context of automated equipment maintenance. When machine learning algorithms are used, the healthy status or the possibility of failure of the equipment is reported in real-time. The historical data is analyzed concerning weather patterns and busy shipping seasons versus slow shipping seasons. Therefore, maritime shipping companies can make adjustments in their vessel voyage plans based on this intelligent information.

Another area where the AI-powered solutions provide immense benefits to maritime shipping companies is in the form of improved safety and security of vessel operations. The AI algorithms do not wait for an attack or a threat to occur and then respond in a reactive mode. The algorithms proactively detect the malicious activities and threats and report to the off shore center so that the ship owners, ship managers, maritime institutes and organizations can take proactive actions and mitigate the risks in the sea environment.

The route optimization can be accomplished effectively by AI-based systems. With the help of these systems, the ship manager can use the most efficient route for the vessel voyage. The prediction of the best path is made by the AI algorithms by considering various key parameters. These include the current weather conditions and minimum fuel consumption.

The vessel performance can also be forecasted by artificial intelligence algorithms. The systems evaluate how significant is the relationship between speed and power and predict changes in the performance of the vessel based on underwater fouling.[li] The historical data is utilized for this purpose to comprehend the rate of degradation that is being faced by a particular vessel.

Another area of AI that has attracted the maritime professionals is known as generative AI.[lii] These AI-based systems possess the capability of generating unique content. This new aspect of AI systems can address various limitations of the current systems and open new opportunities for ship owners, ship managers, maritime institutes and organizations. In this regard, a practical, real-life example is that of Greywing.[liii] It is a company based in Singapore and offers various maritime AI platforms and solutions. The company developed an AI chatbot that is popularly known as SeaGPT. This chatbot was developed based on the recent GPT4 technology. The intent of this solution is to facilitate the communication between port agents and crew members. This tool can be considered as a representative example of generative AI in the maritime shipping sector. The tool automates the whole process of communication. Even the process of drafting emails and sending them are automated. When the information is received by the ship managers from the port agency, the relevant information is extracted by the generative AI algorithms and sent to the specific and concerned member of the crew. It improves the communication process significantly because the information is collected quickly and is also processed accurately.

Figure 36 and Figure 37 show the interface of SeaGPT. By using SeaGPT, the crew members can engage in AI-based chat. They also receive suggestions on the best course of action. The emails can be drafted easily and communication can be managed effectively. The user can enter voice messages as well as send texts. A good aspect of SeaGPT is that only the relevant information is collected that is useful for the algorithms for providing suggestions and recommendations. Therefore, the response time of the chat interface is very quick and prompt. The alerts are also shown on the chat window as shown in Figure 37.

FIGURE 36: MAIN INTERFACE OF SEAGPT[liv]

FIGURE 37: CHAT IN SEAGPT[lv]

This new aspect of generative AI has various use cases that can transform maritime shipping for the future. Its first use is in port operations. The berth planning and traffic management are optimized by using generative AI. The stacking of the containers and the processes of their retrievals can all be optimized. In some advanced implementations, even automated vehicles and cranes can also be used.

The second use case of generative AI is in shipping and logistics. The vessel speed and shipping routes can be optimized. There is an enhanced visibility of the supply chain and the cargo. There are also reduced carbon emissions and fuel consumption. The third important use of generative AI is in maritime safety and security. The algorithms of generative AI can predict the possibility of collisions and accidents with an amazing level of accuracy and precision. The possibilities of smuggling and piracy can also be detected. The illegal fishing activities can also be tracked by the AI algorithms.

Another significant use case of generative AI is the development of drone ships and autonomous vessels. It shows a marked improvement in communication systems and maritime navigation. The control systems are also enabled for voice communication and there is an availability of radio frequency spectrum with an optimized management.

The above use cases of generative AI have made it evident for ship owners, ship managers, maritime institutes and organizations that the future of maritime shipping will be based on those AI technologies that are multi-modal and powered by GPT4 and other such powerful systems. The unique aspect of GPT4 is that the systems not only focus on the content but also on the context. As a result, there is a generation of a more realistic content that could not be produced by the earlier AI systems. Another unique capability of the multi-modal systems is that data can be analyzed and processed from multiple sources that can also include videos and images. The data processing may also facilitate speech recognition. With these feature-rich AI systems and applications, risk assessment and decision-making processes are improved significantly. These AI-based tools and technologies are game-changers for the maritime shipping industry. The ship owners, ship managers, maritime institutes and organizations can be confident regarding profitability, safety, and efficiency.

As there is a significant initial investment required for implementing maritime shipping solutions, I would recommend to ship owners, ship managers, maritime institutes and organizations to do extensive research for selecting an AI company and solution provider. Ship Technology has mentioned the names of the key AI vendors[lvi] that can become a good starting point for you to do your research. Your own research is crucial because no single vendor can satisfy the requirements of all maritime shipping organizations. There will definitely be some solution providers particularly suited for your organization. The vendors included in the recommended list of Ship Technology are Orca AI, Sea Machines Robotics, Buffalo Automation, Bedrock Ocean Exploration, i4 Insight, Ladar, Ocean Infinity, and Massterly.

The shipping industry, as a whole, is going through a phase of digital transformation. The AI is the driving force behind the process of transition. It is because when AI-powered systems are integrated into the shipping industry ecosystem, there is a marked improvement in the shipping operations with respect to sustainability, safety, and efficiency. The vessel operations are the most suitable area of AI intervention because a lot of data has to be processed in real-time for intelligent decision-making. Time becomes a crucial factor in the whole vessel operations. Therefore, the ship owners, ship managers, maritime institutes and organizations will need to evaluate their current business strategies because the only way to move forward with confidence in maritime shipping operations is through incorporating AI. They will have to include AI in energy management, predictive maintenance, safety operations, and cyber security. They will also have to consider the option of autonomous ships to improve the safety and reduce the risk of human error. If you have not implemented AI in maritime shipping business, then you are already behind other successful AI-powered maritime shipping companies.

Kongsberg Maritime is a pioneering example, where the autonomous navigation systems powered by AI were implemented in 2020.[lvii] The successful operations were launched in a ferry in Norway. The system successfully navigated even through complex waterways. The docking procedures were executed successfully without the need of any human intervention. Similarly, predictive maintenance has also been used successfully by various companies. Rolls Royce is a leading example in this regard, where the predictive maintenance system was implemented successfully in 2019.[lviii] Energy management systems have also been deployed successfully by many organizations. These systems can optimize the environmental performance of the vessel operations and can reduce the fuel cost. In this regard, a pioneering example is that of DNV GL. The company successfully installed energy management system powered by AI in 2019.[lix] The system optimizes the power systems and propulsion of the vessel and minimizes carbon emissions and fuel consumption.

Cyber security systems have also been deployed successfully in many organizations. More and more vessels are being operated in a connected environment that has also increased the risk of cyber attacks. AI algorithms can address this issue effectively. A leading example in this regard is that of CyberKeel. The cyber security systems of the company are powered by AI algorithms.[lx] These systems use machine learning techniques to respond to the threats in an automated manner and in real-time during vessel operations. These examples clearly suggest that many organizations including those in maritime shipping have already realized the significance of AI and implemented AI-powered systems. If other ship owners, ship managers, maritime institutes and organizations do not embrace these powerful tools, it will be difficult for them to compete in the maritime shipping sector where there is an intense competition for gaining a market share and tapping new customer segments.

One of the recent trends that should particularly be focused by ship owners, ship managers, maritime institutes and organizations is the use of AI in port operations and cargo handling. Earlier, the focus of the AI systems was on improving the vessel operations. But now, the shipment processes are also being optimized by using automated cargo handling and improving the efficiency of the business processes at ports. In this regard, a notable contribution has been made by APM Terminals where the company introduced a cargo handling system that is powered by AI systems.[lxi] The system uses the algorithms of machine learning and optimizes the processing of containers during port operations. The savings are gained by maritime institutes and organizations in the form of reduced manual labor and improved efficiency.

AI has also been used successfully in the management of supply chain and logistics. The state-of-the-art systems have been introduced by Maersk and IBM. These systems focus on the optimization of cargo handling, route optimization, and scheduling. The systems also minimize the risk of losing a cargo by enabling an effective tracking of the cargo. In this regard, a notable example is that of Watson Supply Chain announced by IBM. This AI module by IBM uses machine learning algorithms to process data from multiple sources and enables real-time decision-making for cargo shipments and route selection.

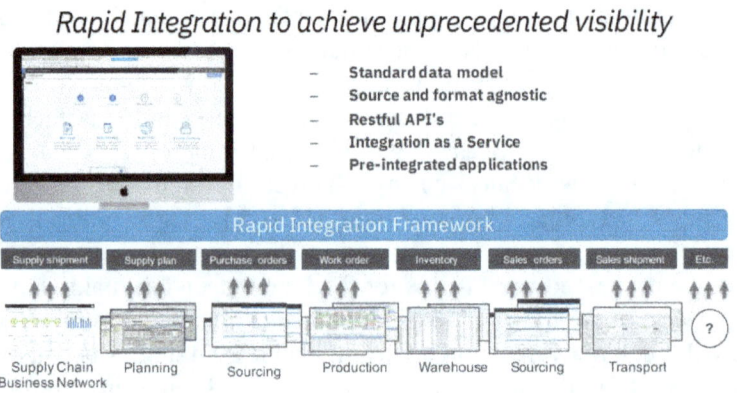

FIGURE 38: SUPPLY CHAIN MANAGEMENT BY IBM

As shown in Figure 38 above, the supply chain management platform of IBM is based on a rapid integration framework. This integrated system enables increased visibility to the business data. The system works on a standardized data model. The system is agnostic to source and format of the data and can process data received from multiple platforms and sources. The system is based on an advanced concept of restful API. From the cloud computing perspective, the system has been implemented based on the model of integration as a service. The system is offered to the end users as a pre-integrated application. The system works seamlessly in all modules from supply shipment to sales shipment. The supply chain business networking is accomplished in supply shipment module.

The planning process based on AI algorithms is executed in supply plan module. The next important step is the development of a purchase order that is considered as the part of the sourcing process. The production process focuses on the work order activities. The warehouse related activities are considered in the inventory module. The sourcing process is considered again in sales order module. The last and the crucial step is the transportation and all the relevant activities are processed in sales shipment module. As is evident from Figure 38, all the modules of the supply chain are tightly integrated and the intelligent information is supplied to the decision-makers after processing the key data by AI algorithms.

All the above benefits mentioned in this section present a good case for implementing AI in maritime shipping for transforming the future of this industry. In the next section, I will present some case studies for AI-powered businesses with illustrations so that you can get a real feel of the true potential of AI in maritime shipping businesses.

2.2. CASE STUDIES OF AI-POWERED MARITIME BUSINESSES

As I keep mentioning in different sections of this book, some of the maritime institutes and organizations have already embraced AI in maritime shipping and it has provided promising results to those organizations. Other maritime businesses will also have to embrace this technology because AI is not only 'enemy' of the conventional jobs but it will also eliminate those businesses that insist on doing businesses in the old fashion. Here I am mentioning case studies of AI-powered maritime systems so that you could realize how AI is transforming maritime shipping for the future.

The first successful case is of Nautilus Labs (NL). NL has optimized the fuel efficiency of the vessel voyage by using AI-based predictive analytics. This AI-based platform makes use of machine learning techniques for evaluating speed, weather conditions, and engine performance. The insights and recommendations by the AI algorithms are used for optimizing the fuel usage. It has resulted in a substantial cost reduction in NL and the company has also been able to improve its environmental track record. A snapshot of the predictive analytics used by NL is shown in Figure 39 below. As is evident from the dashboard, the algorithms consider the voyage history and fuel efficiency is accomplished by considering various internal and external parameters of the voyage operations. Many maritime businesses are using this dashboard for enabling fuel efficiency in the maritime shipping.

FIGURE 39: PREDICTIVE ANALYTICS USED BY NAUTILUS LABS[lxii]

Another renowned vendor offering AI-powered solutions is Shone. The company offers AI-powered systems that optimize the navigation capabilities and the offered systems are also beneficial for autonomous ships. The technology used by Shone is based on sensor fusion, computer vision, and machine learning techniques that ensure the safety of vessel operations.[lxiii] When maritime institutes and organizations implemented AI-based systems offered by Shone, a higher level of autonomy was achieved in vessel operations and the organizations were able to reduce the count of onboard crew.

The third prominent vendor is Orolia where the company offers situational awareness solutions. Remember from my discussion on collision avoidance, I explained there that collisions in vessel occur when the area is congested and the ship captain has a reduced level of situational awareness. This deficiency can easily be overcome by deploying situational awareness solutions such as the one offered by Orolia. The system integrated AI algorithms with satellite data, radar, and AIS. The system detects the safety and security threats and facilitates the captains in making informed decisions. The maritime institutes and organizations that have implemented Orolia's situational awareness have been able to ensure the safety of the vessel voyage and handle the potential threats in the surroundings. As shown in Figure 40 below, the systems designed by Orolia are tightly integrated for the onshore and offshore data centers. The SmartFind collects critical data from the vessel and sends to the vessel monitoring system. The sea conditions are also monitored by a module known as Kannad Solo.

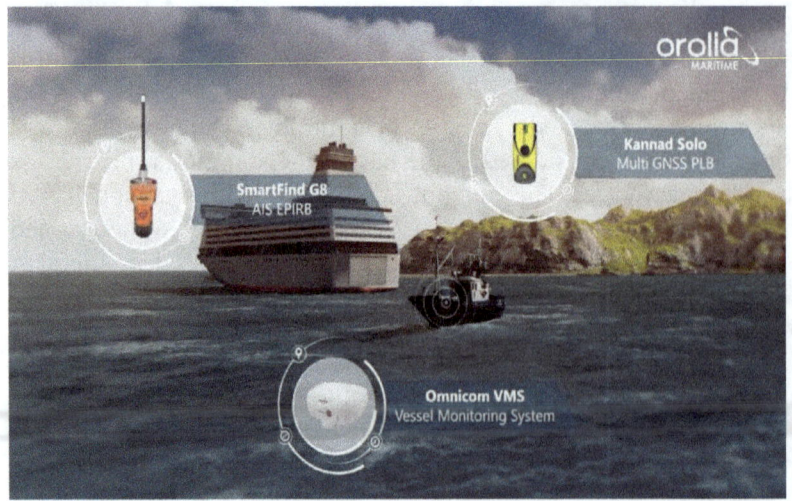

Figure 40: Orolia System for Maritime Securitylxiv

Another notable vendor for maritime businesses is Nautix Technologies (NT). The system makes use of the predictive maintenance feature of the AI. The maintenance algorithms are applied on marine engines and the associated equipment. The system analyzes the historical data as well as the current data from the sensors. The maintenance actions are recommended by the system to avoid equipment failure. Many maritime institutes and organizations have installed NT systems and report significant reductions in downtime and maintenance costs. Therefore, this system should be used by the maritime organizations because it helps them in proactive maintenance planning. As highlighted in Figure 41 below, the solution offered by Nautix is based on offering digital work instructions so that the vessel operations are secure and there is no risk of equipment failure and hazardous operation.

FIGURE 41: SAFETY COMPLIANCE SOLUTION BY NAUTIXlxv

Another prominent vendor for maritime businesses is Voyager Analytics (VA). The company has added a new dimension to AI in the context of maritime shipping. The company offers those solutions that are responsible for the performance and welfare of the crew members. The capabilities of natural language processing are used for assessing the communication among crew members and providing them relevant feedback by the AI algorithms. The recommendations are aimed at improving the job satisfaction of the vessel crew and ensuring their mental health and wellbeing. Maritime companies implementing VA have reported higher job satisfaction of the crew members and the staff turnover rate is also very low in these organizations. As shown in Figure 42 below, VA is one module of a comprehensive platform offered by Voyager Labs and the entire functionality of the system is based on cognitive AI.

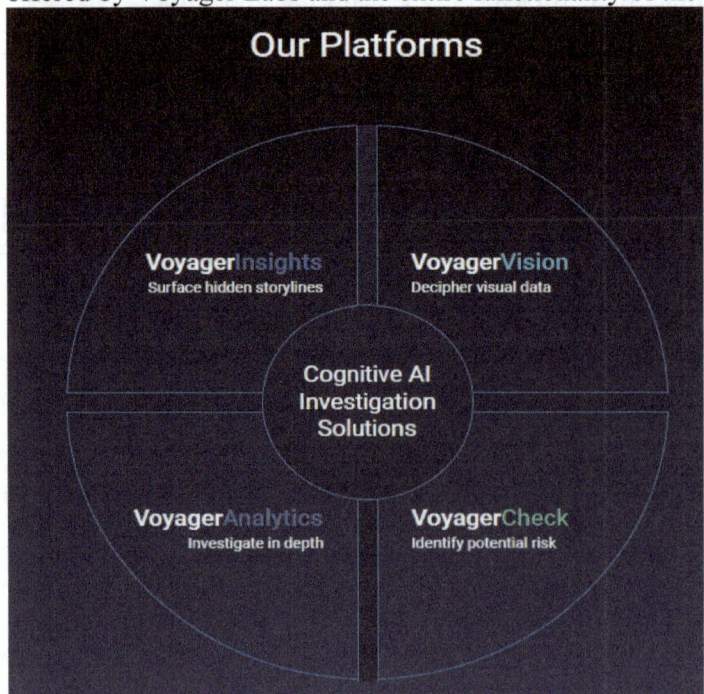

Figure 42: Voyager Analytics based on Cognitive AIlxvi

Voyager Analytics is a recent technology and it is being highly appreciated and lauded in maritime institutes and organizations. It is because the technology focuses on the human factor and aims at improving the job satisfaction of the crew members. While implementing AI, as I mentioned during the discussion of pros and cons of AI, the reliance on machines increases unconsciously and the significance of human power is overlooked. As I highlighted earlier, the AI systems are based on recommendations as well as automation. In the earlier stages, you are most likely to rely on recommendations and leave it to the ship managers to make the final decisions. Therefore, the importance of human capital should never be underestimated because they will continue to play their roles even in AI-powered systems. The technological frameworks have been developed to facilitate the humans and they are not meant to replace them. The human resources will need to upgrade their skills to compete in the new technology-based market, but the role of the maritime businesses should be to assist them in their new journey. The role of the business managers should be that of a mentor so that the motivation level of the staff is at the highest level and their job satisfaction increases with the increase in their service period.

One aspect that I also highlighted while mentioning case studies in the previous chapter that the mentioned vendors are offering AI-based solutions at expensive rates. Therefore, a critical cost-benefit analysis should be conducted before finalizing a vendor. I have mentioned five different vendors in the above case studies. There will also be other vendors that offer their products to ship owners, ship managers, maritime institutes and organizations. The business managers should realize that each organization has its own setup and organizational dynamics. As I recommended, the organizations should also hire AI researchers so that they could give their expert advice after an extensive research which vendor will be the best for a given organization.

It is important to note that the maritime shipping companies will have to acquire solutions from one of the vendors that are offering solutions in their geographical locations. It is not possible for the maritime shipping companies to build AI-based systems on their own. These systems are based on highly sophisticated technologies that cannot be learned easily by maritime shipping companies. IT skillset is a completely different domain and these systems can best be developed by software houses. The maritime institutes and organizations should prefer the acquisition of those systems that are based on global standards and the best practices.

Windward is another organization whose product offerings are based on maritime intelligence. The company offers AI-based products in maritime shipping to reduce the voyage risks and optimize the operational cost. The AI platform offered by the company collects data from multiple sources such as vessel movements, weather stations, and maritime incident reports. Based on the current and historical data analysis, the AI algorithms provide insights into potential opportunities and the risks associated with maritime shipping. The platform can also be used to identify those delays that occurred particularly due to extreme weather conditions. In this manner, the risk of hijacking of ships is also minimized. The latest offering of Windward is Ocean Freight Visibility (OFV). Global freight forwarding companies are using this product developed by Windward and these companies include Metro Shipping, DSV, and Cargo Amerford.[lxvii] The use of Windward solutions by leading maritime shipping companies indicate that ship owners, ship managers, maritime institutes and organizations will have to be proactive in AI-based implementations. If they do not intervene on the patterns similar to the competitors, their businesses will not be able to attract a large pool of customers.

The AI solution improves the scalability and efficiency of maritime shipping by enabling automated data gathering and analysis. The generated reports provide useful and key indicators that include risk predictions, ETA with accuracy and precision, location-based insights, and causes of delay. These are key pieces of information for the decision-makers in the maritime shipping. The customers are highly concerned about ETA in the maritime operations because the order is placed several days before and an expected date is communicated to the customers. If there are further delays, then it causes frustration and anger among the customers. The expectations of the customers can be managed in a better way if they can be made aware regarding the causes of the delay such as extreme weather conditions or unfavorable conditions in a particular area.

The system facilitates the users to input the details of their containers and minimize the risks in maritime voyage. Those disruptions are also reported by the system, which are affecting the ETA of the container. Any evolving situations are reported by the system in real-time. The solution offered by Windward also offers an API integration through which the functionality of the system can be integrated with the current business applications and transportation management systems. It is a highly useful functionality because most of the maritime shipping companies will already be using several software and web-based applications. It will take some time to replace all of these applications with AI-based systems. Therefore, during the initial period of AI implementation, it is a good feature to integrate the existing systems with the AI-based systems.

Figure 43 below shows the dashboard of AI-based OFV system offered by Windward. The system provides port insights as well as shipment tracking. There are three separate tabs of shipment tracking, dashboard, and port insights. On the left side, ETAs are being shown based on maritime AI prediction. The causes of delay are also being reported such as the demurrage risk. As highlighted in the snapshot, the overall performance of the crew members is also indicated as to how many shipments were completed early and on time. It is also shown how many shipments were delayed. The risks in the voyage route are also highlighted in real-time.

FIGURE 43: DASHBOARD OF OFV SYSTEM BY WINDWARDlxviii

Figure 44 below highlights that OFV is just one of the numerous AI-based solutions offered by WindWard. Windward solutions are broadly classified into trading and shipping solutions, supply chain and logistics solutions, and government and public sector solutions. OFV is one of the solutions in supply chains and logistics suite. Other solutions in this suite are know the vessel ETA, predict ETA in maritime operations, ports insights, terminal insights, and exception management. The key solutions in trading and shipping suite are sanction compliance, business intelligence, fuel consumption optimization, and exposing fraud risks/TMBL.

FIGURE 44: SOLUTIONS OFFERED BY WINDWARD IN MARITIME AIlxix

Another useful application that has been adopted successfully by various ship owners, ship managers, maritime institutes and organizations is MarineTraffic. This application is available as a web application and also as a mobile app. The application gives real-time data regarding the vessel movements to ship owners, ship managers, maritime institutes and organizations. The AI algorithms collect data from various sources including coastal radar stations and AIS transceivers. The system is so powerful and intelligent that the algorithms track the movement and location of the ships across the globe. This system is not just meant for ship managers but it is also being used extensively by other stakeholders including port operators and insurance companies working in the maritime sector.

The services offered by the application are paid services and they are offered in four plans of Standard, Professional, Professional Plus, and Global Satellite. When you acquire any of these plans, you will receive a huge volume of data that will facilitate your voyage operations and maritime shipping. As shown in Figure 45 below, different data points provided by MarineTraffic include voyage data, satellite tracking, vessel events, container tracking, weather maps, port congestion, live maps, density maps, and vessel particulars. The availability of the options varies from package to package. The least expensive package is Standard (Column 1) where some of the functionalities are not available. The most expensive package is Global Satellite (Column 4) where the full functionality is available.

FIGURE 45: FEATURES AVAILABLE IN DIFFERENT PLANS OF MARINETRAFFIClxx

Another notable solution provider for AI in maritime businesses is Spire Global. The company is focused on space data collection and analytics. The company uses state-of-the-art AI technology and evaluates data from nano-satellites' constellation. The online platform has proved its significance in maritime shipping and provides insights into key vessel operations including vessel tracking, safety issues, and oil spill tracking. As indicated in Figure 46 below, Spire offers a range of AI-based solutions for maritime shipping. The key maritime solutions include the analysis of historical AIS data, port events analysis, the validation of AIS position, ETA data for vessel to port, and generating the Ship View. The company also offers solutions regarding weather analysis, aviation, space, earth intelligence, and radio frequency intelligence.

FIGURE 46: AI-BASED SOLUTIONS OFFERED BY SPIRElxxi

Another vendor that has gained prominence recently is Sea-Kit International (SKI). SKI is headquartered in the UK and manufactures those marine systems that can be operated autonomously. The systems developed by SKI are useful in various areas of maritime shipping such as environmental monitoring and rescue operations. Environmental monitoring enables the ship managers to schedule the cargo shipments at the appropriate time slots. The rescue operations can be launched effectively by using the AI-powered, autonomous ships of SKI. Therefore, these systems reduce the overall operational cost and optimize the safety of the vessel voyage. A model of the autonomous ship designed by SKI is shown in Figure 47 below. The company has highlighted that using its autonomous vessels provide key benefits to ship owners, ship managers, maritime institutes and organizations.

FIGURE 47: MODEL OF AUTONOMOUS SHIP BY SKI[lxxii]

The first benefit is flexible design. The payload area has been designed to be adaptable. The vessel has been developed to have multi-mission capability. The vessel can be operated as a standalone ship or it can also be made part of a fleet. The vessel is highly customizable and modifications can be made to suit a particular journey and execute specific tasks. The second benefit is cost-effectiveness. Unlike the general perception, it is claimed by SKI that the vessel incurs lower capital expenditures and there are also reduced running costs. The mobilization of the vessel is also accomplished at a lower cost. There is also a higher endurance level of the autonomous vessel. Due to the automated operations of the vessel, the staff cost is also reduced.

The third benefit is the reduced carbon emissions. The ship travels through the sea in a hybrid power model enabling lower carbon emissions. The fuel consumption is also reduced because the vessel uses the optimized route recommended by the AI algorithms. A quiet mode is achieved in the acoustic environment due to the autonomous operations. The vessel also possesses solar charging options and wind charging options that improves the contribution of renewable energy sources in the overall system. The fourth and the last benefit is the improved level of safety. The vessel is controlled remotely with the analysis of all past and current data indicators. The risks and dangers to the human lives are avoided because there is no staff on board. The points of failures are avoided in their entirety due to AI-based predictive maintenance. The design of the vessel is certified by the renowned certification bodies such as DNV-GL.

Maritime shipping companies should also evaluate this model of autonomous vessels. It will enable them to operate in sensitive and war-prone areas as well. The organizations may not risk sending the crew members in those sensitive areas. However, the autonomous vehicles may be sent to gain a competitive advantage and expand the businesses in untapped markets. Drone deliveries are already being considered for the speedy and secure deliveries. So, there is no harm if maritime shipping companies also capitalize on the opportunities offered by the autonomous systems.

2.3. OPTIMIZING LOGISTICS THROUGH AI

In the modern system of trade and commerce, logistics can be considered as the backbone of maritime shipping operations. A proper storage and movement of cargo requires a careful planning and execution. The competitive advantage for maritime institutes and organizations lie in the factors of accuracy, speed, and efficiency, and all these variables can be controlled efficiently by AI-based systems. The integration of AI in maritime logistics can bring a revolution in this domain and the maritime institutes and organizations can receive the benefits of optimized operations. In the era of covid-19 pandemic, the AI-based solutions gained a huge acceptance because robots handled the containers and the shipment process efficiently without human intervention. At that time, there was an overall emphasis on a reduced level of human interaction and adopting social distancing. Therefore, AI-based logistic systems can prove to be handy even in testing times and tough conditions.

AI is optimizing various aspects and factors of maritime logistics. Its key interventions can be seen in the effective inventory management and demand forecasting. The algorithms evaluate the historical data of sales and forecast the demand of maritime shipping with accuracy and precision. AI tools can also highlight market trends for intelligent decision-making. When the companies adopt this forecasting approach, maritime shipping companies can avoid stockouts and minimize carrying costs. Another key intervention of AI is in efficient delivery planning and route optimization. The scheduling and routing of vessel voyage are optimized by AI algorithms where the system considers the parameters of delivery window, vessel capacity, weather conditions and traffic situation.

In the domain of logistics, AI algorithms not only recommend the optimized routes but route adjustments can also be made in real-time in an automated manner. This strategy makes it possible that the delivery of the package is made efficiently and timely. The maritime institutes and organizations achieve significant savings in transportation costs and fuel consumption.

Warehouse operations are also being automated these days by using AI-driven robots. These robots make significant optimizations in warehouse operations. They can retrieve the products quickly, manage the inventory on their own, and package the products for the delivery. It also improves the layouts of warehouses for their efficient utilization. The robots also recommend the optimal storage positions and minimize the time required for picking up the goods and packing them.

Another key intervention of AI is in the optimization of last-mile delivery. In the context of maritime shipping, the last mile of the delivery operation is considered a highly costly step. The ship owners, ship managers, maritime institutes and organizations have to consider traffic patterns, the locations of delivery, and the details of the package. When the parcel reaches the destination port, the work of the maritime shipping does not end. The parcel then has to be delivered to the location of the customer. This task is highly challenging because now the execution is away from the sea route and road transportation has to be used.

The latest technology in this regard is autonomous vehicles and drone deliveries in which AI tools guide the delivery process. These options are being explored for making further improvisations in the last-mile deliveries. Maritime shipping companies should also use those tracking systems that are powered by AI tools. It is because the entire supply chain has an improved and real-time visibility by using these systems. The systems provide status and generate alerts from the procurement stage to the delivery stage. All the stakeholders can identify the movement of the cargo in real-time. Therefore, there is a higher level of transparency and the decision-making is based on the intelligent analysis of data. Figure 48 shows a real-time implementation of AI in maritime logistics. The mentioned solution has been implemented by P&O Maritime Logistics. The company has optimized the offshore logistics by using the AI-based systems. By using the AI algorithms, the cargo delivery operations were executed with reduced number of vessels, cargo runs, and carbon emissions.

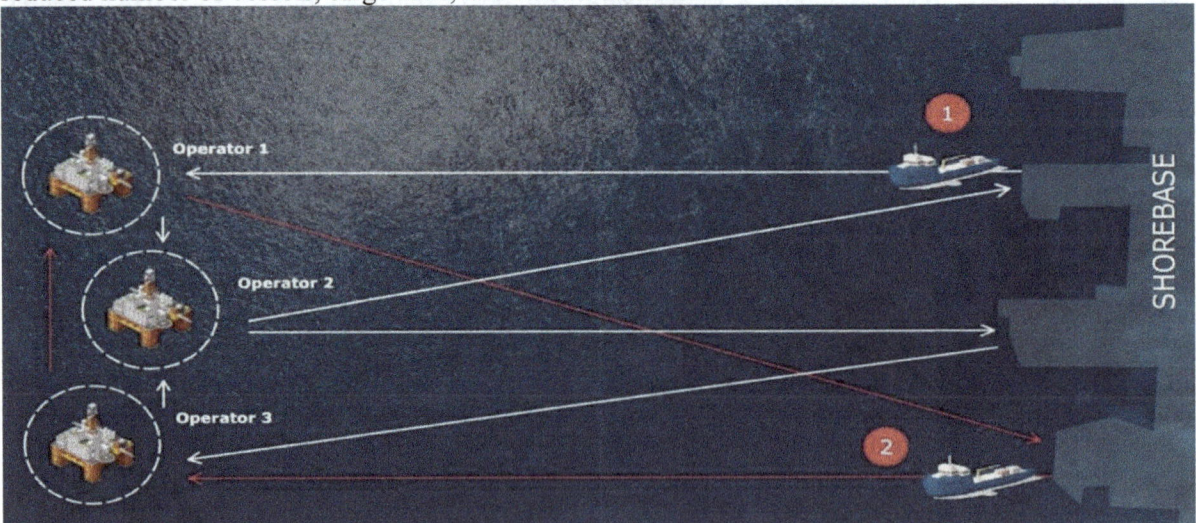

FIGURE 48: AI-BASED LOGISTICS IMPLEMENTED BY MARITIME LOGISTICSlxxiii

P&O Maritime Logistics achieved route optimization highlighted by the red line. It shows that various touchpoints were avoided and reduced by using an AI-based system. The figure also shows that the two vessels were operating in the sea and the AI-based systems enabled a coordination mechanism between these two vessels. The need of operator 2 was eliminated due to the optimized operations.

Customer orientation and customer-centric approach also play a pivotal role in the domain of logistics. AI technology has become quite mature in this area and virtual assistants and chatbots are being used in various industrial organizations and sectors. The customer service is enhanced significantly by using these AI tools. By using these tools, orders can be tracked, inquiries can be handled, and valuable information can be made available to the customers. A good aspect of these AI tools is that chatbots and virtual assistants are available for the customer service 24/7. The personalized assistance offered by these tools improves customer engagement, customer satisfaction, and customer conversion.

AI tools also facilitate in load planning and freight selection. The algorithms assess the nature of goods, available options of carrier, and shipping requirements and enable a cost-effective transportation. The systems can also recommend the most suitable load configurations and carriers. Maritime shipping operations are quite expensive and each vessel voyage incurs a huge cost. Therefore, fraud detection and risk management are considered crucial activities in maritime shipping. AI algorithms assess the potential risks to the operations in real-time. They also assist ship owners, ship managers, maritime institutes and organizations to identify the potential disruptions in real-time.

Vessel operations are not only challenged by extreme weather conditions but they may also be challenged by political instability and unexpected attacks during the sea travel. AI-based systems can also identify anomalies in maritime shipping transactions so that the business operations are not affected by fraudulent activities in the payment systems and the supply chain.

Due to the numerous technical resources available to ship owners, ship managers, maritime institutes and organizations, AI revolution has also encouraged the initiatives in maritime shipping regarding green logistics and sustainable logistics. The lower carbon emissions improve the emission record of the maritime organization and fuel consumption is also optimized. The transportation options adopted by ship owners and ship managers are always eco-friendly and sustainable packaging practices can also be used. All these interventions can transform the future of maritime shipping by using AI-based tools, technologies, and systems. The companies can drive sustainability, increase operational efficiency, and minimize the cost of operations.

2.4. OPTIMIZING SUPPLY CHAIN THROUGH AI

When we talk about a modern supply chain in the maritime shipping, it is more than just the acquisition of an order to the delivery of the package. It is now an interconnected network in which multiple processes and transactions are involved and multiple stakeholders participate in strengthening the supply chain management. Maritime businesses should use AI-based systems for optimizing the supply chain management. These systems can entirely transform the way supply chain management is currently being practiced in maritime institutes and organizations. The AI algorithms can introduce predictive analytics, real-time insights, and automation in the supply chain.

The most critical factor in the supply chain of maritime shipping is the relationship optimization and supplier management. The AI algorithms evaluate the performance of a supplier based on the historical data and they consider the factors of costs, quality of service, and delivery times. As a result, the most reliable suppliers can be selected. If the organization wants, the other suppliers may also be communicated regarding their reasons of rejection by the AI algorithms so that they could improve their services for a future selection. AI-based systems also improve the collaboration and relationship with the suppliers and streamline the entire process of procurement.

Another important area where the AI facilitates in supply chain is the assessment of the financial stability of the suppliers. The business risk is reduced by evaluating environmental risks and geopolitical factors. The potential risks are identified by using predictive analytics and maritime businesses are facilitated in enabling smooth operations and producing contingency plans. Figure 49 below shows one model of an automated, AI-based supply chain in which the robots are being used for transferring the cargo from the containers to the ship. The benefits are achieved in this AI-based setup in the form of faster transportation, reduced costs, lesser fuel emissions, seamless inventory tracking, and a significantly higher level of productivity.

FIGURE 49: AI-BASED MARITIME SUPPLY CHAINlxxiv

2.5. INTELLIGENT DECISION-MAKING WITH AI

Decision-making is one area where the significance of AI cannot be more emphasized. The decision-makers in maritime shipping include ship owners, ship managers, maritime institutes and organizations. Any flaws in the decision-making on their part can incur significant losses to the organization. Moreover, there is a huge inflow of data from the sea environment, shore data center, and internal organizational setup. It is humanly impossible to process and analyze this voluminous data. Only AI tools and technologies possess the capabilities to process and mine big data and predict the key parameters with precision and accuracy.

AI can transform decision-making in maritime shipping in multiple areas and fronts. The first key area is the processing and analysis of data. The AI algorithms can process both structured data and unstructured data. You will be surprised by the unprecedented speed of AI systems in processing big data. The decision-makers then gain valuable insights and find the patterns in data that enable intelligent decision-making. Machine learning techniques form an integral component of the overall AI tools. These techniques uncover hidden patterns in the data and then predictive analytics can guide ship owners, ship managers, maritime institutes and organizations.

Another important area of intervention is the forecasting tools. In maritime shipping, the environmental variables have a higher level of uncertainty and the captains have to make key decisions while the vessel is travelling in the sea. The predictive models of AI facilitate the tasks of ship owners and ship managers because the future trends are forecasted based on the analysis of historical data. These tools also assist in predicting customer behavior and enabling proactivity in the decision-making process.

One area where the project managers and ship owners have to focus particularly is the resource allocation and resource optimization. The AI algorithms have been proved to be highly effective in resource allocation and their recommendations are spread on all areas including human resources, budgeting, and time management. The optimization models enable maximum productivity and indicate the best strategies for human resource optimization, fuel optimization, route optimization, and time to delivery optimization.

Another key dimension that is of crucial significance for decision-makers in maritime shipping is the management and mitigation of risk. The AI algorithms evaluate the possible risks and quantify the severity level of each risk. They evaluate historical data as well as the risk factors in the current environment. The ship owners, ship managers, maritime institutes and organizations can then formulate effective risk management strategies. The algorithms even suggest how the decision-makers can mitigate those risks based on the analysis of incident reports, global trends, and the best practices.

Through the machine learning models, the risk management strategies of the decision-makers are always optimal and these models also alert the management regarding the unusual patterns in the vessel movement.[lxxv] Therefore, the decision-makers can develop proactive strategies to mitigate the potential risks.

The conventional decision support systems used by the ship owners and ship managers mostly offer generic tools and interfaces and there are no personalized recommendations based on the profile of the individual and the organizational setup. When the AI-based decision support systems are used, the recommendations are always personalized and customized. The insights provided by the algorithms are governed by the historical interactions, behavioral patterns, and preferences of the individual. When the ship owners, ship managers, maritime institutes and organizations receive relevant and customized information, their choices and selections are more intelligent and informed.

AI-based decision support systems are also preferred due to the fact that the information and insights are provided in real-time. The decision-makers can then respond promptly to the changing circumstances in the vessel operations and can also respond to the evolving market dynamics. The processing of live data can be done efficiently by AI-powered systems and time-sensitive and critical decision-making can be optimized by AI tools.

In the era of AI-based environments and organizational setups, the decision-makers in the maritime shipping are also opting for automated decision processes. It means that the decision-making should not be slowed down by red-tapism. The decision-makers just provide the broad guidelines and the ground level decisions are taken automatically by the AI-based systems. The decision-making models can work in AI tools with an amazing level of accuracy and precision. By using the predefined parameters and rules, the AI models can ease the cognitive load of ship owners, ship managers, maritime institutes and organizations and they can allocate more time to strategic decision-making.

A new trend in implementing AI-based systems is to use 'explainable AI'.[lxxvi] These systems improve the trust level of the decision-makers on AI tools and technologies. As I explained to you at various points in this book, the AI-based algorithms process a huge amount of current and historical data. They also use their training and learning to arrive at the recommendations, predictions, and solutions. However, the decision-makers have to present a strong case to the senior management and the management may also ask the rationale and the logic behind selecting a particular course of action. Therefore, the decision-makers should also know how the AI models arrived at a particular solution or a recommendation. The Explainable AI is the solution to this requirement where the AI-based systems also provide insights regarding the working of the AI algorithms in a simple and easy-to-understand language. It creates a high level of transparency in the decision-making process and the recommended solution can be justified easily to all the stakeholders.

Another unique aspect of AI-based systems in the domain of decision-making is that the systems do not work in a reactive mode but believe in continuous improvement and learning. So, let's say, the AI algorithm recommended a solution and you implemented it in maritime shipping. The solution proved 80% significant but there were limitations and issues in this solution to the extent of 20%. When you supply this information to the AI systems, a process of learning and improvement will take place. The next time, when you ask for a decision from the AI-based system, the system will also consider the limitations in the previously recommended solution and give you the best solution by considering all the constraints. This process of learning and adaptation enhances the decision-making potential of ship owners, ship managers, maritime institutes and organizations. Moreover, the AI models remain effective and relevant because the improvements in the systems keep them up-to-date and efficient.

2.6. OPTIMIZING SAFETY AND SECURITY WITH AI

The vessel voyage is full of challenges, hazards, and risks. The risk factors are higher than air travel or road travel. There have been incidents where the entire ship was damaged and the maritime organization was no more sustainable. Therefore, the optimization of safety and security is always a high-priority item for ship owners and ship managers. The maritime institutes and organizations will be highly pleased by the fact that AI tools have opened immense opportunities for optimizing safety and security in maritime shipping.

When AI-powered systems are used, the surveillance and monitoring of the vessel voyage can be accomplished in real-time. These systems use computer vision to evaluate the video feeds of the vessel operations. The suspicious behavior and unauthorized accesses are immediately reported by the system. The threats are detected by the system by processing vast amounts of data. The systems not only focus on the sea operations but also evaluate public safety concerns and cyber threats. Machine learning techniques provide information regarding the suspicious network activity and data security of the connected environment is also ensured.

Various touchpoints in a large vessel can also be controlled by AI-based systems of biometric recognition. The finger prints and facial patterns can easily be recognized by these systems for the identification of the crew members and access control. The access may be denied to certain privileged areas based on the access role matrix. Another useful AI-based intervention in the security domain is known as predictive policing.[lxxvii] In this approach, the AI algorithms evaluate the historical data gathered from law enforcement agencies. This data is used for predicting the likely occurrence of a crime at a given point in time of vessel voyage. The law enforcement agencies can then be informed accordingly by the ship owners and ship managers so that they could allocate personnel on high-risk areas. The criminal activities in the sea area can be minimized significantly by using this approach.

Figure 50 shows an implementation of AI-based maritime security system. As highlighted, the vessel management does not have to focus exclusively on the own vessel operations. Instead, the security system should consider the whole maritime commons. The security of the surface vehicles as well as undersea vehicles should be evaluated by the systems. The key aspects of the AI-based maritime modules include the port surveillance systems, ISR, electronic warfare devices, and the support from the special operation forces. This security system has been adopted by the defense department of the United States. It indicates the usefulness and relevance of AI-based maritime security.

FIGURE 50: AI-BASED MARITIME SECURITY[lxxviii]

Another potential safety hazard in the vessel operations is the fire hazard or a natural disaster. The vessel carries valuable items of the customers. Therefore, the loss will not only be limited to the belongings of the maritime organization but the damages and penalties may also be suffered in the form of the clients' claims. The AI-based systems can monitor environmental conditions far better than the conventional systems. The hazards relating to fire or natural disasters are reported and alerted well before allowing an adequate time to the decision-makers to respond to those hazards.

The algorithms can also respond by using sprinkler systems. The emergency responses may also be initiated by the system so that human lives could be protected and damage to the valuables could be reduced. However, as I explained in the previous sections, it is up to the decision-makers of maritime shipping to use AI-based systems only for recommendations or also for automated responses. The safety and security is one area where I would highly recommend to use automation in AI-based systems. It is because in the case of fire or natural disasters, the time to respond is too limited and a systematic response is needed to prevent the losses at a mass scale.

Another area of concern for ship owners, ship managers, maritime institutes and organizations is the safety and security of financial data. The AI algorithms can also detect unusual behaviors of users while executing payment transactions. The fraudulent activities can be tracked instantaneously and the alerts are generated in real-time. If the automated response is activated in the AI-based systems, the payment transactions may even be declined in those cases. The maritime businesses can enable corrective and preventive actions in the context of financial frauds by using AI.

Another crucial aspect in the maritime shipping operations is the protection of critical infrastructure. The vessel movements are facilitated by state-of-the-art transportation systems, power grids, and water facilities. If a security breach occurs, the organizations may face a huge loss to their critical infrastructure. Therefore, AI-driven monitoring should be enabled in maritime businesses. The essential services should run smoothly and they should have an all-time availability through a strong monitoring system.

As I highlighted in the above paragraphs, safety and security of the vessel operations is a critical success factor and it is high time for the ship owners and ship managers to incorporate AI in the security paradigm. The protection of the assets and individuals will become increasingly challenging by using

48

the conventional tools, and only AI-based systems can respond to the emerging threats in the on-shore and off-shore environments.

2.7. AI FOR CREW MANAGEMENT

Effective crew management can be a game changer in maritime shipping because the quality of the human resources defines the quality of the services. At one hand, the maritime businesses should optimize the crew operations, and on the other hand, the ship owners and ship managers should also work on the welfare and benefits of the staff. In all these aspects, the use of AI can revolutionize the entire crew management.

The planning and scheduling of the staff can be optimized by AI algorithms. The systems analyze the skills level and preferences of the staff and assign them the duties accordingly. The algorithms also ensure the compliance of the HR business processes with the state and federal labor laws and regulations. As I mentioned to you in the earlier sections, when the AI-based systems are introduced in maritime institutes and organizations, there is a need for training the crew for using these new systems. The AI experts will just install and configure these systems and then it is up to the crew to make the most of these new systems. The systems are beneficial only up to the extent to which the intelligent reports are utilized by the crew and the management.

If the automated response is not enabled and the crew also ignores the alerts and recommendations made by the AI systems, then these systems will lose their significance. Even at the managerial level, I explained that now there is a trend of using Explainable AI. Even the managers struggle in comprehending the reports generated by the AI systems. Therefore, before and after the implementation of AI systems, the business managers will have to focus on the skill development and training of the crew. AI-based systems can also be used for this purpose. The algorithms evaluate the current skill set of the crew members based on the assigned tasks and job responsibilities.

Machine learning techniques are then used to recommend customized training programs for the crew members. The training and skills development of the crew members is essential because it will improve their performance in using AI-powered systems and also enhance the safety awareness of the crew.

Another crucial aspect of crew management is providing feedback and a continued performance monitoring. In the AI paradigm, it can be accomplished by collecting data from various sources instead of relying on a single source or one individual. The automated performance tracking systems can evaluate the performance based on the feedback of supervisors, customers, as well as peers. Various vendors also offer AI-based modules known as performance analytics.[lxxix] These systems assist the management in identifying areas of improvement for an employee and give them relevant and timely feedback.

As I mentioned earlier, crew management is not all about monitoring the performance of the staff during the vessel operations. The health and wellbeing of crew members is of equal significance and advanced AI tools should also be used for the health monitoring of the staff. Different AI gadgets such as sensors and wearable devices can be used for this purpose. The ship owners and ship managers should track various key indicators of the crew members that can affect their performance levels during the voyages such as stress levels, fatigue level, and whether proper sleep is being taken regularly. When the health issues of crew members are identified at an early stage, it will be possible to make timely interventions. AI tools should be used for promoting the physical and mental health of the crew so that medical emergencies could be avoided during vessel operations.

Another area of consideration in the crew management is to allocate adequate work hours for the crew members so that they also have some time for the rest. However, ship owners, ship managers, maritime institutes and organizations operate in a highly competitive market of maritime shipping. They also need to ensure that the order completion and delivery rate is 100% and the staff is demonstrating optimal productivity during the work hours. The AI algorithms can assist in this regard and optimized work hours and rest hours can be set for the crew. The algorithms will make recommendations based on the nature of the vessel operations, work schedules, and regulatory requirements set by the state authorities. The maritime businesses should also note that their work schedules are compliant with the international regulations because the maritime shipping operations are considered as transnational operations. Therefore, the AI algorithms should be trained to enforce strict compliance with the frameworks such as Maritime Labor Convention.[lxxx] The communication and collaboration among the crew members can also be enriched by using AI tools. When the communication platforms are powered by AI tools, there is a higher level of information sharing and cooperation among the crew members. AI tools become particularly relevant when the teams contain members from different countries and cultural backgrounds. AI tools offer the features of real-time language translation due to which language barriers are removed and effective communication is facilitated.

In the context of crew management, the AI tools are also highly effective in emergency responses and crisis management. The systems recommend optimal response strategies to save the lives of the crew members and protect the valuables from destruction. The AI tools are also powerful in generating scenarios and simulations. Through these simulated environments, the crew members can be trained for dealing with emergency situations.

The leave management and the daily roster of the crew can also be automated by AI. The algorithms evaluate the preferences of the crew members, and at the same time, they also analyze the operational requirements. After evaluating all the factors, the best leave approval strategy is suggested. As a result, there is always an adequate staffing for managing the vessel operations.

While the AI-based systems for crew management can provide extensive functionality on their own, they can also be integrated with the organizational HR systems and crew portals. As a result, there is an integrated platform available for fetching all the information related to the crew. AI systems also offer a self-service option whereby the crew members can view and even update their non-critical information such as change of address or phone number. These aspects not only optimize the HR functions but also improve the level of transparency in HR operations.

As you have already seen and appreciated that there are various areas of crew management where the business processes can be transformed by unleashing AI. It is often a neglected area of AI implementation because ship owners, ship managers, maritime institutes and organizations are more focused on the core business operations and the significance of a quality human resource is often overlooked. However, the effectiveness and efficiency of maritime shipping operations are highly reliant on a good health and wellbeing of the crew members. When the crew members are highly satisfied with their jobs, there will be a lower turnover rate and AI systems will be used by the crew for optimizing all vessel operations and shipping processes.

2.8. AI FOR PREVENTING CYBER-ATTACKS AND FRAUDS

Every intervention in the maritime shipping business provides new opportunities and also poses several risks and challenges. The AI implementation is no exception and various challenges may be faced when the system is fully implemented and functional. In the AI-based systems, there is a lot of interconnectivity among the devices, applications, and tools because the data from multiple sources is considered by the AI algorithms for decision-making. When there is to and from movement of the data from various sources, the dataset may reach the unintended recipients or the information may also be hacked by the criminals as a planned cyber attack. The developers of AI-based systems have also anticipated these attacks and AI-based systems have also been developed to prevent frauds and cyber attacks.

AI-powered systems evaluate the user behaviors in the interconnected systems and also assess the overall network traffic. The anomalies and unusual patterns in the network traffic are immediately reported to the data centers. The machine learning techniques continuously evolve them to combat the recent strategies of the hackers. The AI algorithms also detect if there is any malware injected in those systems that are using AI tools. The presence of malicious software, ransomware, and viruses all are detected by the algorithms.[lxxxi] The threats are blocked before the hackers are successful in their motives. AI-based antivirus solutions are also being offered currently that makes it possible to fight the threats through the real-time protection.

In the context of maritime shipping, access management and identity management are highly significant for ship owners and ship managers. It is because all crew members are not authorized to visit all parts of the vessel. Moreover, there have been incidents where the individuals were on-board the ship without the approval because they wanted to leave their countries and reach a safe destination. These aspects can affect the reputation of maritime institutes and organizations and the unauthorized personnel can also damage the property and belongings. When AI-based systems are used, the security and access mechanisms are optimized by using behavioral analysis and biometric verification. The automated systems reject the request to access if the analysis concludes that it is an unauthorized entry. Within these systems, the multi-factor authentication can also be enabled so that the identity of each user could be verified through multiple mediums.

The response against the cyber attacks should be automated by AI tools. It will ensure that the impact of the attack is minimal and the response is quick. Moreover, as I explained to you at various stages, the AI algorithms also learn from the historical data. Therefore, the response will also train the algorithms and the subsequent responses will be more effective and targeted.

As you can see from my discussion above, the cyber attacks are one of the biggest challenges in the interconnected environment of AI systems. The AI tools should be used by ship owners, ship managers, maritime institutes and organizations for coordinated responses and automated actions. A secure digital ecosystem is critical for gaining the maximum benefits from AI-powered systems.

2.9. AI FOR A BETTER CUSTOMER SERVICE

Customer orientation and a customer-centric approach are highly recommended in the business setting, and maritime shipping is no exception. There has been commendable work done so far in AI-based development concerning customer service. The virtual assistants and AI-based chatbots have been welcomed by all industries and sectors. Below, I will explain to you how AI can transform the maritime shipping in the domain of customer service.

Customers now have higher expectations from the maritime organizations and they want the availability of the customer support round the clock. Earlier, it was challenging for the organizations to enable customer interactions all the time. However, virtual assistants and chatbots have enabled the business entities to process a high volume of customer queries round the clock. The good aspect of these AI-based technologies is that they are expert in the processing of natural language such as the chatbots. The bots can understand the customer queries and concerns and respond in a way similar to humans. In this way, the response time improves significantly and the accuracy level of the provided information is also improved. In the case of a human assistant, the individual may forget a point and provide wrong information. However, chatbots respond based on the data fed to them. The humans may also get frustrated and angry with the aggressive remarks of the customer, whereas the chatbots listen to all the responses without anger. The customers may also find it convenient to talk to bots because they can talk with an open heart without a fear that a given comment will be disliked by the opposite party.

AI-based systems and services also provide customized recommendations to the customers based on their past visits to the website and purchase history. The promotions and recommendations regarding the maritime shipping services are always based on the preferences of the visitor. This aspect of personalization improves customer satisfaction and customer engagement.

A prominent area where the AI systems gained a huge prominence in the initial days of AI-based tools was the sentiment analysis.[lxxxii] The user generated content on social media and blogging websites have become a valuable asset for the maritime shipping companies. They want to analyze the customer feedback and sentiment to improve their services and know the positioning of their services. The sentiment analysis can only be made by AI-based algorithms because the traditional computer programs and software rely heavily on quantitative data. The intelligence of natural language processing is rare in these programs, and therefore, AI-based tools should always be used for sentiment analysis.

The predictive analytics powered by AI can also be used effectively in customer support. The analytics tools make predictions regarding the possible customer issues and customer needs. As a result, the customer support team can become proactive in providing support to the valuable customers.

AI-powered tools are not only effective in the processing of natural language but they can also work efficiently in speech recognition and voice recognition. The voice-enabled interactions are highly demanded by maritime shipping customers and AI-based tools can also recognize a customer based on their voice.

A key area of customer service is conflict management and complaint resolution. When there is a huge inflow of customer complaints, ship owners and ship managers are always worried how to resolve them and improve the maritime shipping process so that the complaints could be reduced. The AI-powered systems also assist in complaint resolution by recommending the solutions based on the analysis of the past data. The algorithms evaluate how the similar complaints were resolved earlier successful with a good customer feedback. The same solution is suggested to the customer support personnel. As the AI-based technologies are efficient in natural language processing, the customer feedback and complaints are processed in real-time and they are categorized based on the analysis of comments and sentiment analysis.

You will have realized by now that the AI systems have a huge potential in optimizing the customer service in maritime shipping. From the personalized recommendations to the suggestions for complaint resolutions, the role of AI is there, at every touchpoint. The ship owners, ship managers, maritime institutes and organizations should make the customer service based on AI technology to increase the customer conversions and customer retention.

In this chapter, my focus was on highlighting how AI is beneficial in every aspect of maritime business. I highlighted its benefits and use cases in various business units, departments, and areas of business. As the saying goes, 'all that glitters is not gold'. The same is the case with the AI tools and technologies. It will not be prudent if you blindly follow the AI path in maritime shipping without realizing the initial investment involved and the challenges associated with AI implementation. Although the focus of this book is on highlighting how AI can transform the future of maritime shipping, I feel it my responsibility to also make you aware regarding the challenges with AI implementation. When you have the idea of both benefits and challenges, you can make a cost-benefit analysis in your specific organizational context as to whether go for AI implementation or not. I am still confident that your final decision will be to go for AI implementation because the benefits of AI implementation in maritime shipping significantly outweigh its cost and challenges.

In the next chapter, my focus will not be exclusively on highlighting technical challenges in AI implementation. As the AI is a new paradigm in the technological world, the regulatory frameworks are also updated time to time and cultural challenges also emerge in different regions and geographical locations. To know about all these aspects, let's proceed to the next chapter.

3. CHALLENGES WITH AI IMPLEMENTATION

I have mentioned innumerable benefits of implementing AI in maritime shipping for transforming its future. However, similar to other technology experts, I am not oblivion to the challenges and issues in its implementation. It's a two-edge sword for ship owners, ship managers, maritime institutes and organizations. At one hand, they cannot do away with AI implementation because other maritime shipping companies are using AI for improving and optimizing their business processes. But they have to justify this huge investment to the senior management. Therefore, they need to have a sound understanding of the challenges and issues involved in implementing AI in maritime shipping.

Another aspect that I have focused in this chapter is to make you realize that the AI implementation challenges are not just technical challenges. An aspect of novelty is associated with AI implementation not only for ship owners and ship managers but the state and federal regulators are also struggling in responding to the AI revolution. There is so much data exposure to the AI-based systems that the security and safety of the sensitive data is at stake. Therefore, regulatory frameworks are being evolved from time to time. It can pose serious challenges to the implementation because once the implementation is finalized, it will be difficult to modify the AI paradigms for the alignment with the new regulatory frameworks.

Another challenge that I have focused in this chapter is the cultural challenges. AI has also generated a fear factor in societies because most of the tasks that humans have boasted that only they can execute them as experts can now be done easily by AI systems. Just think about the autonomous ships in the maritime shipping. If it is implementing in a maritime shipping, then the jobs of most of the crew members are at stake. Think of route optimization techniques done by AI algorithms, then the jobs of route planners are in danger. Think of tools such as SeaGPT. The jobs of manual communicators are gone. So, how can a ship manager implement AI technology successfully in maritime shipping that could be a win-win situation for all? It should be a source of satisfaction not just for external customers such as clients and suppliers but also for internal customers such as employees. When ChatGPT was introduced, the jobs of content writers and marketing professionals were at stake. When robots will carry out the work of cargo handling and container handling, what will the ground staff do? How the technology can be implemented without affecting the interests of the current workforce? These are some of the interesting topics and issues that I will cover in this chapter.

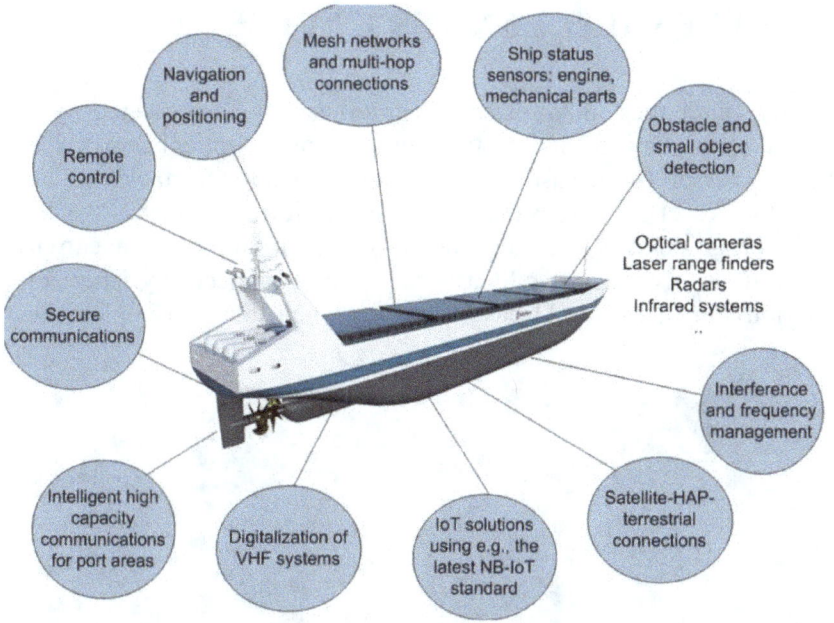

3.1. TECHNICAL CHALLENGES FOR THE MARITIME INDUSTRY

One of the most significant areas where you will have to focus your attention while implementing AI is the technical challenges. Strategic solutions are required for addressing the technical challenges in the maritime shipping because the challenges are intricate and some of these challenges are specific to maritime shipping.

One of the topmost challenges that may be faced in the maritime shipping sector is the availability and quality of data. The AI algorithms are only as intelligent and competent as the training data available to them. If you cannot provide a good quality of data in a sufficient volume to the algorithms, the results produced by AI algorithms will not be promising.

As an example, try using ChatGPT in a language other than English. Open AI claims that the system can support multiple languages, but the results in other languages may not be as meaningful as in the English language. It is because there is no person sitting on other side of the chat. There is a chatbot on the other side that is expert in natural language processing. But if the data is not available sufficiently on the web in a particular language, what the AI algorithms can do? The quality of the results will deteriorate significantly in other languages.

The same is the case with AI algorithms for maritime shipping. These algorithms have a huge reliance on extensive, high-quality, and diverse datasets. Only through quality datasets, the AI algorithms can perform the process of training and improve the precision and accuracy of the models. Currently, the datasets available to maritime shipping for historical analysis are either in unstructured format or are inconsistent. It may greatly affect the accuracy and the development process of AI models.

Another technical challenge that might be observed in maritime shipping is the level of standardization and integration achieved in the underlying data. Various systems and sensors provide data to maritime shipping systems that are powered by AI. The data might be collected from the weather stations, the vessel itself, the logistics chains, and the ports. It is a really complex and challenging task for the AI algorithms to standardize and integrate this data into AI systems. There is a need of achieving standardization in maritime-based AI systems so that formats and protocols could be decided for the AI systems. Then all the information sources should provide data in the same format to enable effective analysis in AI algorithms. Currently, this level of standardization is not available because the maritime institutes and organizations are competing in AI implementations for gaining a competitive advantage. They should build collaborations so that a standard data format could be agreed upon at the industry level.

Another technical issue that may arise in the maritime shipping is in the context of connectivity and edge computing. Maritime vessel operations are all about operations in the far flung areas of the sea where there may be a low connectivity in a remote area. When the AI-powered systems are used, all the connectivity and interfacing among various systems is achieved by the internet cloud. The real-time transmission of data may become highly challenging and complex when a good level of internet connectivity is not available for transmitting the data to onshore AI systems.

FIGURE 51: CONNECTIVITY CHALLENGES IN VESSEL OPERATIONSlxxxiii

Figure 51 highlights the connectivity challenges in the vessel operations that are powered by AI systems. The connectivity is needed for various instruments and data providers. Optical cameras and radar systems are also connected to the vessel. In the case of autonomous ships, the obstacles are also detected by internet connectivity and the vessel operations may be hazardous if the connectivity is lost. Similarly, the navigation and positioning systems are also dependent on a good internet connectivity. The IoT solutions in the vessel operations are also heavily reliant on the internet connectivity. Therefore, a lower connectivity can pose a serious threat to the AI-based systems when they are implemented extensively in the port operations.

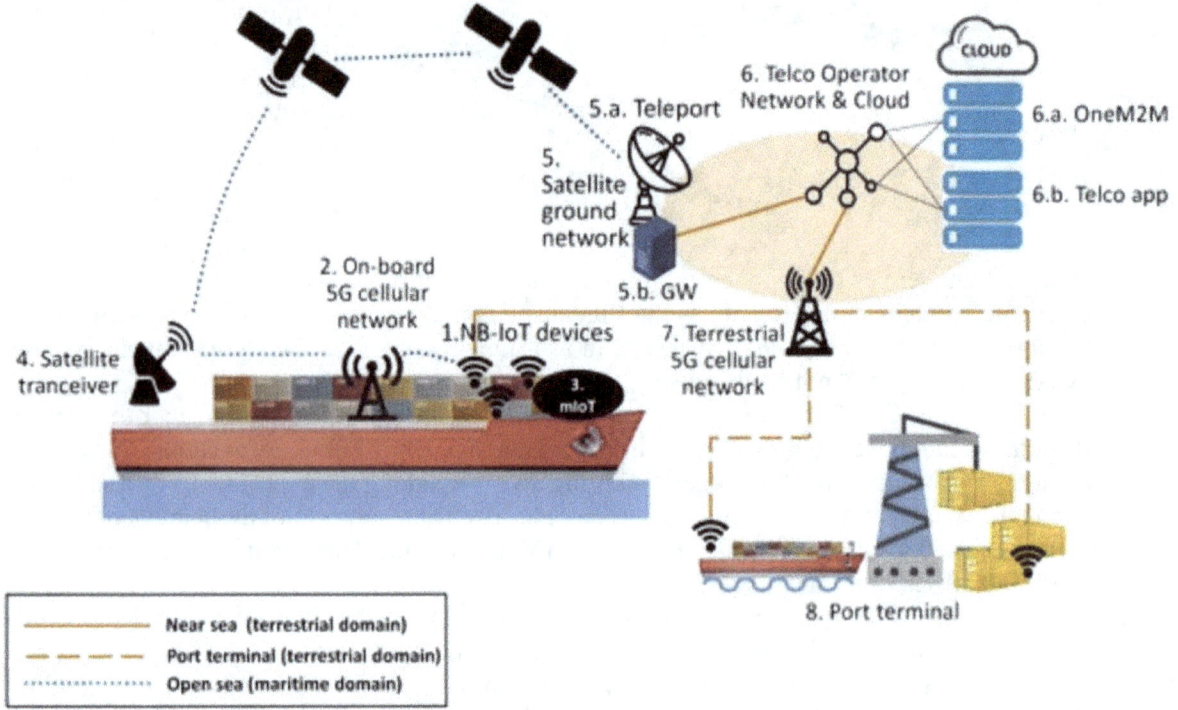

FIGURE 52: IOT BASED MARITIME SHIPPING SERVICESlxxxiv

Figure 52 also shows a promising IoT-based, AI model of maritime shipping services. The model is based on 5G cellular network and automated operations have been enabled at near sea domain, port terminal, and open sea domain. As is evident from the figure, the high-bandwidth internet connectivity is needed at all touchpoints through the traditional internet service providers as well as satellite-based service providers. There could have been significant service disruptions in the vessel operations if a high-level internet connectivity is not enabled in this system.

Another issue arises in the context of edge computing where the AI implementation becomes complex at the edge of onboard vessels. It might occur because the limitations are associated with the hardware configurations. In those cases, the AI models might be burdened to operate with minimum computing resources and minimum availability of data that will reflect badly on the quality of the results and recommendations.

There is also one more technical issue associated with AI implementation in maritime shipping. An extremely high computational power is required for the training of AI algorithms. It is because the training dataset is usually very huge and the algorithm has to 'save' all the combinations for utilizing them in the suggestions, computations, and recommendations. The maritime shipping industry is more focused on the sea environment and the shipment processes. Allocating such a high computational resources in the corporate environment and the cloud computing environment can become a real technical challenge for ship owners, ship managers, maritime institutes and organizations.

Another unique aspect of the AI models is that for them, learning is a continuous process. The performance of an AI model today will not be exactly the same after several days or months. The algorithms retrain themselves and are focused on continuous updating. This adaptation to the changing maritime requirements and conditions will require huge computational power whose provision can become challenging for the IT team of the maritime shipping company.

One more challenge in the AI implementation that will be faced inevitably is the privacy concerns and the issue of cyber security. For hackers and cyber criminals, maritime industry is an attractive target because the IT expertise of the crew members are limited and the stolen data is highly valuable in terms of monetary value. The hackers can also exploit potential vulnerabilities in the vessel operations that may emerge due to the integration of the IT systems. The ship owners, ship managers, maritime institutes and organizations will have to implement strict controls and measures to prevent data breaches and unauthorized accesses.

One challenge that the maritime shipping companies are bound to face in the AI implementation is that they will need to integrate the AI systems with the existing legacy systems. It is almost impossible for any maritime shipping company to transform all its current systems to AI-based systems in one go. It will definitely be a gradual and evolutionary process in which the AI-based technologies will be introduced in different domains periodically. In chapter 3, I have mentioned various areas where maritime AI can be introduced in the businesses. These areas include logistics, supply chain, decision-making, crew management, cyber security, and a better customer service. So, suppose you purchase IBM's AI-based logistics system. Now, you will have to integrate other systems supply chain, crew management, customer service, and so forth with the AI-based system. It will be a real technical challenge because the legacy systems might be based on very old technologies, and compatibility issues may be experienced when integrating these systems with AI technologies.

In the previous chapter, I also talked about Explainable AI. It is a highly demanded feature by business professionals because they do not just want a rapid, intelligent report generated by the AI system. They also want to know how the algorithm worked and arrived at a particular solution. This functionality is available in Explainable AI but all AI-based systems do not offer this functionality. Therefore, ship owners, ship managers, maritime institutes and organizations may struggle in comprehending the reports generated by the AI systems.

Most of the AI-based algorithms are based on deep learning models. Therefore, they are popularly known as black boxes. They are termed it so because the algorithms are highly complex and it becomes difficult for the business managers to know the rationale of the decision-making done by AI algorithms. The ship owners and ship managers will have to ensure interpretability in AI algorithms during the implementation. This aspect is particularly crucial in safety-related domains so that trust and compliance could be built regarding the AI-based systems.

Another issue that may arise during the AI implementation is the issue of cost and resource allocation. AI implementation is a time-intensive and resource-intensive activity. The ship owners, ship managers, maritime institutes and organizations will have to allocate adequate resources for the implementation. This can be quite challenging in the current context when the organizations are already burdened financially to retain their existing workforce. The existing workforce is already involved in executing daily maritime shipping operations. When they are asked to devote their time for learning and implementing AI-based systems as well, the resistance might be shown by the existing workforce.

The technical resources will also be needed such as the server machines, licensed software, and skilled IT professionals. Therefore, the implementation will not be materialized if the senior management is unwilling to allocate sufficient financial resources. Due to these aspects, the smaller maritime shipping companies might find it difficult to implement AI-based systems. They should collaborate with big shipping companies and replicate their AI-based scenarios in their organizational setup so that the outcomes could be achieved with minimal resources and the investment of time. The ship owners and ship managers will have to make key decisions where the cost-effectiveness should be balanced by the allocation of adequate resources for a successful AI-based implementation.

Another issue that may arise in the maritime shipping industry is the complex and global nature of maritime shipping operations. A wide range of vessels operate in the sea environment. I mentioned to you several case studies where the maritime shipping industry has also embraced AI. However, these are only a few examples. In the other cases, the maritime shipping industry is still relying on conventional systems. Therefore, it becomes highly complex to develop AI models in these environments such that the models could be applied to all dimensions of the industry. For example, the collisions are accidents can be avoided if AI systems have the knowledge of the surrounding vessels and they will predict the risk of collisions based on their learning and the operational framework. However, if the surrounding vessels are not AI-powered and your vessel is the only vessel powered by AI, the usefulness of your systems will be limited. AI is all about gathering data from multiple sources and working in an integrated, connected environment. If that level of integration is not available, the benefits of AI-powered systems might be diluted to ship owners, ship managers, maritime institutes and organizations.

Based on the above issues, the technology professionals should follow an evolutionary approach to introduce AI-powered systems in maritime shipping industry. First, they should start at a small level. It would not be prudent to automate all maritime shipping operations in a single go. The technology managers should identify a few tasks in the business where the automation can bring significant improvements by using AI. Second, the IT managers should define different phases for implementation. Initially, the pilot projects should be started for AI implementation, and then the AI operations should be scaled up as the IT team, crew members, and other staff to gain more and more understanding of using AI-based systems in maritime shipping.

The technology professionals should also recommend to the senior management to make investments in data quality. As I explained earlier, the effectiveness of AI algorithms is heavily dependent on the quality of data. Think of an AI chatbot where a user asks it to develop a CV for him. The user provides details regarding qualifications and experience. If the user provides incorrect details such as mentioning qualification as PhD incorrectly, and mentioning experience of 20 years as teacher incorrectly, the algorithm will not be able to detect the false assertions by the user. The CV will be developed and curated based on the description provided by the user. The same is that case with the algorithms related to maritime shipping. If the algorithms do not get the data of good quality, the vessel movements are reported incorrectly, the weather conditions are reported with inaccuracies, then the recommendations and suggestions of the AI algorithms will be based on this data and it will be meaningless for ship owners, ship managers, maritime institutes and organizations. Therefore, a heavy investment should be made not only for configuring AI-based systems but also on the acquisition of high-quality data.

Figure 53 below shows how the machine learning algorithms work and how they can go wrong if the quality of data is not impressive. Machine learning algorithms enable the high power computer machines to decide on their own without the manual programming instructions of a computer programmer. For this purpose, the machine needs the training data. A comprehension of the data is developed with the help of the AI algorithm. When a machine develops a good level of understanding of the data, it develops a decision-making model based on the training data. Then these decisions are not recommended instantaneously but are first tested for accuracy. It makes a continuous loop in the AI systems and a high level of confidence is achieved by feeding more and more training data. Therefore, the confidence level of the machine learning algorithms depends on how good is the quality of data provided to them. The figure also highlights that the actual data or the problem is dealt by the model that has already developed a good learning by the training data. This model is then used for making predictions.

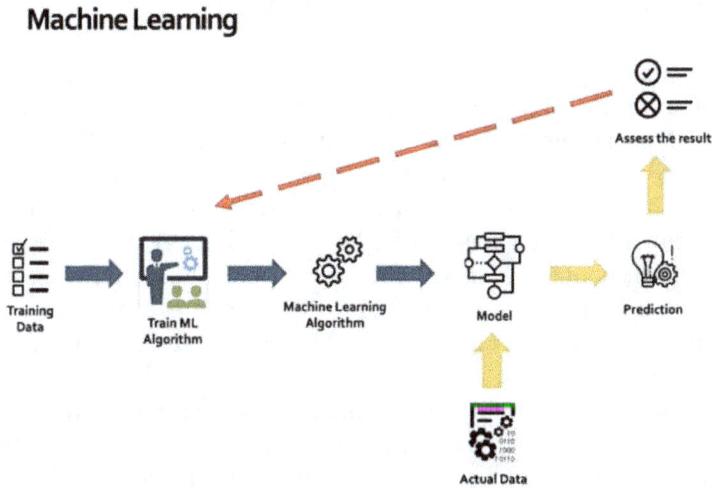

Machine Learning

Training Data → Train ML Algorithm → Machine Learning Algorithm → Model → Prediction → Assess the result

Actual Data

Figure 53: How Machine Learning Algorithms Work[lxxxv]

From the above technical challenges, it is very much evident that ship owners, ship managers, maritime institutes and organizations will have to think deep in technical terms before implementing AI-based systems. Some of the ship owners and ship managers might realize that this level of sophisticated technical expertise is beyond their capacities. In those cases, it is better to build partnership with experts. As I have explained and mentioned, various maritime shipping companies have successfully implemented AI-based systems. Those success stories can be replicated in your organization or their expert team can be asked for facilitating in the implementation.

I mentioned earlier that one of the technical challenges in the AI-based implementation in maritime shipping is that the adoption level of AI is very limited and the real benefits of AI-based systems can only be achieved if all the companies develop a maritime shipping AI ecosystem. Therefore, if you try to build partnerships with other organizations in the maritime shipping industry, there is a great chance that they will welcome your partnership because an increased level of AI adoption is also in their own interest.

Some AI experts have also warned that an increased level of reliance on AI-based systems can increase the risks in the defense and security systems. These aspects should also be focused by the maritime shipping industry. There have been cases recently where the fighter groups penetrated the maritime regions of the adversary countries and the radar systems of the country could not detect this penetration. The AI experts have explained that it occurred due to the low data quality available to the AI algorithms. The intelligence agencies of the country reduced their own surveillance and intervention and instead began to rely on the AI systems for smart surveillance. However, the enemies were successful in affecting the quality of data that was being supplied to the machine learning algorithms.

As an example, suppose an AI-based system generates risks and alerts in the maritime region based on a sudden or unusual movement of vessels in their sea area. However, if an enemy falsely feeds the data to AI system that there is no such unusual movement and the vessel operations are within the acceptable limit, then the machine learning algorithms will not generate any alert. As I explained in Figure 53, the system uses the model to make predictions and generate alerts and the model is totally reliant on the data fed to it. Therefore, these risks should also be assessed by ship owners, ship managers, maritime institutes and organizations.

Based on the above scenarios, I would highly recommend that the deployed systems should undergo a rigorous level of testing before implementation. Moreover, for a certain period such as three to six months, the conventional systems should also be run in parallel. When the ship owners and ship managers have a complete confidence on the AI-based systems regarding the prediction accuracy and alert system, only then these systems should be made fully operational and conventional systems should be stopped.

APAC Entrepreneur has mentioned three major technical challenges in AI implementation and these challenges are also significant for maritime shipping industry.[lxxxvi] The first challenge might be faced by ship owners and ship managers when there are inaccuracies in the data analysis by the AI algorithms. In the conventional computer systems, there is a popular phrase that if there is garbage in, there will be garbage out. The same is also true for AI algorithms. The AI programs can learn only those pieces of information that are provided to them. They do not possess 'intuition' or any self-learning mechanism. If the administrators of the AI-based programs provide unreliable or incomplete data to the AI systems, there might be skewed, biased, and inaccurate results produced by the algorithms. The quality of data governs the smartness of an AI program.

An interesting example has been quoted in this regard by APAC Entrepreneur. Amazon has always welcomed the use of AI in its different business units and departments. In 2014, the AI programs were integrated into the business processes of the HR department. The training data was supplied by Amazon for the learning of AI algorithms. The data consisted of resumes and job applications submitted to Amazon during the past 10 years. As the participation of women in the workforce is increasing at a gradual pace, the majority of the job applications in the database were by males. The AI system then committed error in its analysis and incorrectly concluded that being male is a preferred quality of a candidate to be hired at Amazon. As a result, the AI program rejected many applications submitted by the female candidates.[lxxxvii] This example indicates that if the data is skewed to a particular attribute, the whole working of the AI algorithms can go wrong.

The second technical challenge highlighted by APAC Entrepreneur is the algorithmic bias of the AI systems. Although AI systems work in an automated manner and the training and learning mechanism is built into the system, the system is first developed by a human AI programmer. Therefore, there is a significant risk that the AI systems might be biased or faulty, and it all depends on the technical skills of the AI programmer. The faulty algorithms cannot be relied upon and they may give you unreliable results in critical situations in maritime shipping. A bias occurs in an AI-based program when a programming strategy is favorable to a self-serving criteria. The algorithmic bias has even been observed at the level of search engines and social networking sites. Black Entrepreneur reported one such example of an AI algorithm that Facebook developed to prevent hate speech. The program was developed in 2017. However, the algorithm removed the posts related to hate speech only in those cases where the hate speech targeted white men. The algorithm did not remove posts when the hate speech targeted black children.[lxxxviii] Later, it was found that the algorithm had not developed categories precisely due to which hate speeches related to black children could not be identified by the algorithm. It could have been a lack of skills of the programmer or a biased approach on the part of the AI developer. Therefore, the development of AI algorithms should also be reviewed critically concerning the algorithmic bias.

The third challenge highlighted by APAC Entrepreneur is the black box nature of the AI algorithms. When the AI systems make recommendations and suggestions, it is usually an outcome of processing a huge volume of data. The data-driven decision making looks a promising initiative, but what if a ship owner or a ship manager is asked to provide the rationale for the decision-making. In most of the cases, the ship owners, ship managers, maritime institutes and organizations will have to develop the rationale on their own because it will be difficult for them to know how the AI programs reached at a particular decision. It is a big drawback of the system because you are not just interested in knowing the right path. You are also and even more interested in knowing why a given path is better than all other paths. In the AI world, finding those answers is complex if not impossible.

Another technical challenge in the implementation of AI-based systems has been highlighted by Dr. Rory Hopcraft.[lxxxix] He is a renowned cyber security lecturer at a renowned institution. Dr. Rory explained that the vessel ecosystem is full of those ships that can be termed as older ships. In an interconnected and cloud-based AI environment, these vessels become the weakest links. As there are not latest systems and software installed on those ships, the other systems also become susceptible to cyber attacks and malware attacks.

All these technical challenges make one aspect very clear. The ship owners, ship managers, maritime institutes and organizations are domain experts and they are usually not experts in AI implementation. So, they have to rely on the expertise of others for AI implementation. It is creating a low pace of adoption of AI systems particularly in the maritime shipping. The stakes are very high, and if the implementation has technical flaws, the ship owners and ship managers will be the first that will be blamed by the organization for their hasty decisions.

Considering the above technical challenges, it is always prudent to get advice from the technology professionals, maritime shipping professionals, and business leaders before implementing a solution in your company. The pros and cons of all available solutions should be evaluated critically. You will have to adopt the AI route sooner or later, so I would highly recommend to evaluate more and more solutions before it is too late and you lose your valuable customers and clients.

As I said earlier, the description of these technical challenges is in no way meant to discourage you from implementing AI in maritime shipping companies. This description is aimed at making you realize that AI-based implementation is not a bed of roses. You will face challenges, and if you know about these challenges, you can be more proactive in addressing those challenges. These challenges can be addressed by a collaborative effort. You should involve all the key stakeholders such as regulatory agencies, technology professionals, and maritime shipping professionals.

3.2. REGULATORY CHALLENGES FOR THE MARITIME INDUSTRY

You should not expect that AI implementation will just be a 'plug and play' and will integrate smoothly in the current regulatory framework. When the benefits of the AI implementation will be realized by ship owners, ship managers, maritime institutes and organizations, the regulators will also raise their eyebrows because too much data exposure will be seen in AI algorithm generated reports. As the maritime vessel operations are transnational operations, the situation becomes further complex because the ship managers may think erroneously that they have cleared all the regulatory requirements and all of a sudden, there are new regulations in a new territory.

The regulatory challenges in AI implementation need to be examined carefully by ship owners and ship managers because they have far-reaching implications. They might even result in completely abandoning the AI-based system because as I explained to you, the performance of the AI-based systems is reliant on the quality of the data and the availability of data. If the restrictions are imposed by the regulators on the availability of certain data points, the performance of the AI-based systems will degrade substantially. In the worst case scenario, these systems might work like conventional and ordinary systems in the absence of data because there is not a process of learning and adaptation.

The first and the biggest regulatory challenge in the AI-based implementations is the adherence of the maritime shipping companies with the maritime regulations. Different regulatory frameworks are applicable to maritime institutes and organizations. An international body has also been established that oversees the operations of maritime companies and this body is known as IMO (International Maritime Organization).[xc] The maritime shipping companies have to obey the standards set by IMO. It is a challenge for the AI developers to develop and integrate AI solutions in maritime shipping companies that are compliant with the international codes and the regulatory frameworks such as International Safety Management (ISM) and Safety of Life at Sea (SOLAS).[xci]

Another regulatory challenge that makes the AI implementation in the maritime shipping industry a daunting task is the lack of AI-based regulations. It keeps the senior management always worried that if there are new regulations introduced in the future, their whole investment might go wasted. The regulatory grey area is a big concern for ship owners, ship managers, maritime institutes and organizations because it shows an element of uncertainty. When there are restrictions imposed on the acquisition of data due to the regulatory frameworks, it becomes difficult for the AI algorithms to maintain their accuracy and precision.

There is one more challenge in the regulatory paradigm and it is related to security compliance and data privacy. When the AI algorithms use personal and sensitive information, compliance issues may emerge. Therefore, the ship owners and ship managers should also take a legal advice regarding the data privacy laws. In this regard, the internationally known regulatory framework is General Data Protection Regulation (GDPR).[xcii] Google has also issued an alert to its publishers that they should mention on their websites that their site complies with the clauses of GDPR. In the case of AI-based systems, the compliance with the GDPR becomes more complex due to the fact that most of the processes in the AI-based systems are automated. Moreover, if the business managers are not using Explainable AI, they often do not know if the AI-based systems are following the GDPR clauses or not. Although GDPR regulations have been formed by the European Union, but its compliance requirements are universal. The maritime operations are in any case transnational, therefore, the ship owners, ship managers, maritime institutes and organizations will have to ensure that their AI-based implementation is complied with the GDPR clauses.

GDPR is a comprehensive regulatory framework regarding the data protection. Its articles are mentioned in a total of 11 chapters as shown in Figure 54 below.[xciii] The first chapter discusses the general provisions of data protection. The second chapter presents the key principles of data protection. The chapter highlights that the data access should be consensual and data processing should be carried out by considering different categories of personal data. The third chapter highlights the rights from the perspective of the data subject. In this chapter, article 22 also discusses the notion of automated decision-making and profiling that is typically carried out by AI algorithms. The working of those AI algorithms should be complied with article 22.

FIGURE 54: GDPR 11 CHAPTERS

The fourth chapter discusses the role of the controller and the processor. The controller has some general obligations mentioned in section 1. The controller also has obligations regarding the security of personal data, and these obligations are highlighted in section 2. In this regard, it has been made mandatory to the organizations that if a personal data breach occurs, it should not be hidden intentionally by the organization. Instead, a notification should be sent to the supervisory authority. The data breach should also be communicated to the data subject.

Chapter 5 contains those articles that are related to data transfer when international organizations or third parties are involved. It is the case in the maritime shipping operations because various third parties are involved in the supply chain and the data regarding weather patterns, fishing activities, and vessel movements are gathered and transferred to the international organizations by the AI-based systems. Therefore, these systems should be complied with the provisions mentioned in chapter 5.

Another important chapter in GDPR in the context of maritime shipping is chapter 8 that mentions penalties and liabilities. This chapter clearly highlights the regulatory challenges in the implementation of AI-based systems. If these systems are not complied with the regulatory frameworks, the maritime institutes and organizations may face heavy penalties. The data subjects have various rights to liability and compensation that have been mentioned under article 82. All these articles highlight the regulatory challenges that might be faced in the implementation of AI in maritime shipping.

Another regulatory challenge in the AI-based implementation is that the accountability and the determination of liability become extremely complex. As I explained to you, you will not be the only manager or the organization that will be implementing AI in the maritime shipping. Many organizations have already embraced AI and all these implementations create an AI-based ecosystem.

In this AI-based ecosystem, AI-related incident may occur due to the negligence of any one maritime organization. However, since there is so much interconnectivity established among the AI-based systems, it will be difficult to hold any single entity accountable for the incident. The critical maritime decisions may go wrong because of the negligence on the part of the AI developers in a single maritime organization. However, the whole industry might face the blame of such incidents due to a high level of interconnectivity.

The regulatory bodies will also find it challenging to present a forensic evidence in the court. The working of the AI algorithms is even challenging for an ordinary software developer. The judges in the judiciary cannot be expected to comprehend the working of AI-based algorithms. In these cases, the role of the electronic and social media becomes extremely crucial. The media can play a constructive role in explaining the situation in technical terms, or it may also get involved in propagating fake news. As the truth is only known to the AI experts, spreading fake news is very easy in the AI world, and it also poses a challenge to the regulatory bodies.

There are also ethical issues involved in the use of AI. For example, in most of the countries there exists a competition commission that ensures that all business entities have a level-playing field in the corporate environment. They excel in the market based on their efforts and competence. Let's say there are three maritime shipping organizations working in a country. One of them implements advanced, AI-based systems for maritime shipping. The other two companies will be almost out of the competition and may become bankrupt. Therefore, the growing use of AI is also being seen from a humanitarian perspective.

The safety concerns are also being reported by the data subjects because they are not certain about the confidentiality of their data. The AI algorithms may use their data without their consent. As an example, if you write a prompt on ChatGPT or Google Bard that asks to make a summary of the book, the prompt will be executed and you will get a good level and professional summary of the book. However, that book might be copyrighted and the author might not want to read that book without purchasing it. But the AI algorithms suddenly breach this contract and commit copyright infringements. These types of ethical issues may also be faced in the maritime shipping organizations. Therefore, I have devoted an entire chapter (Chapter 6) in this book to highlight ethical considerations in implementing AI.

There is one more regulatory challenge in AI implementation and that is cross-border regulations. The vessel operations are global operations and the vessels are stationed at different ports. Therefore, the ship managers and owners will have to comply not only with the local regulations but also the regulations of those countries where the cargo is to be delivered. It is difficult to comply with the regulations of all destinations because some countries have adopted a very negative stance regarding AI implementation. For example, some countries have officially banned the use of ChatGPT. Moreover, some countries do not offer the facility to connect and receive data from their satellites. Therefore, while on the one hand, it is the strength of the AI systems that the data is collected and processed from multiple sources and historical data is also evaluated, but on the other hand, it also becomes an issue when the acquisition of data is hurdled by the local regulatory bodies.

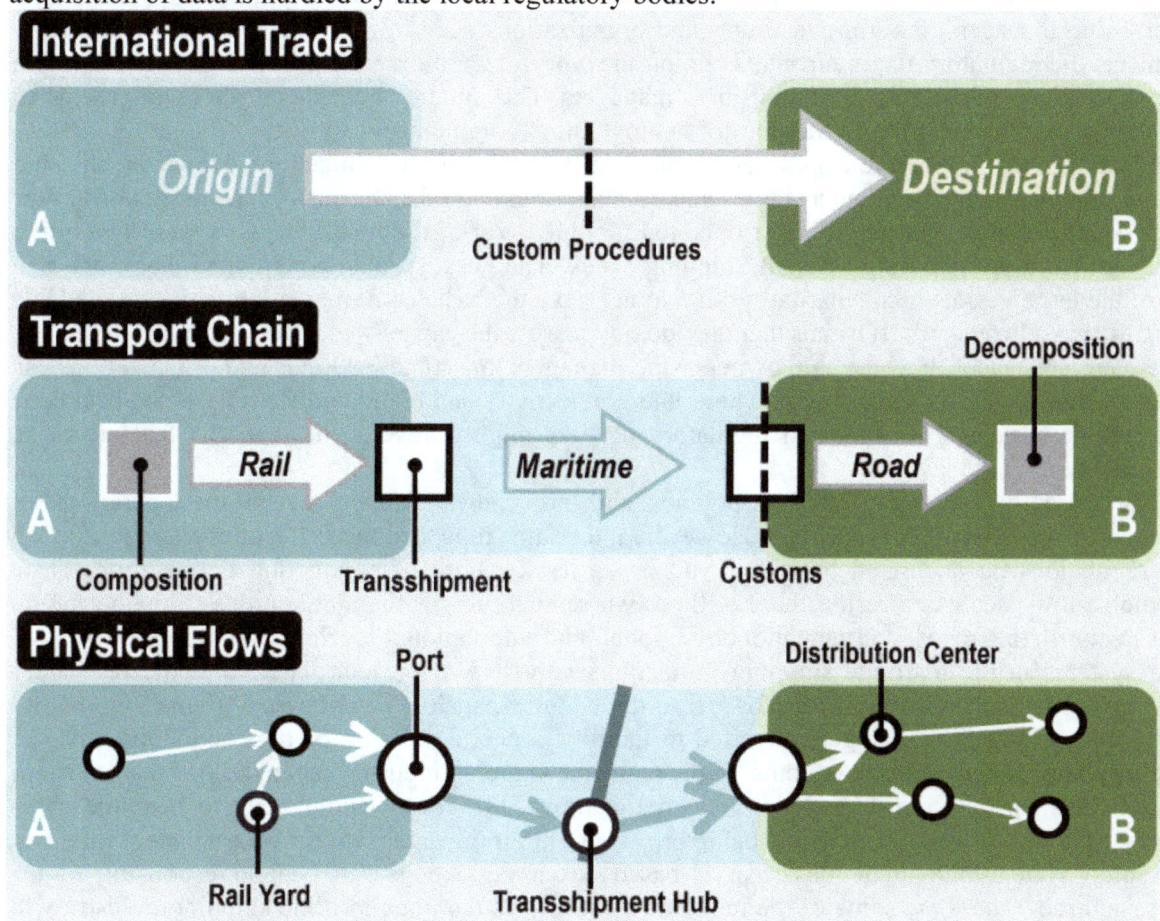

FIGURE 55: TRANSNATIONAL MARITIME SHIPPING OPERATIONSxciv

Figure 55 highlights the transnational nature of maritime shipping operations where the ship owners, ship managers, maritime institutes and organizations have to consider the regulatory frameworks of multiple destinations and geographic locations. The first lens through which the maritime shipping operations should be seen is the lens of international trade. From this perspective, the maritime organizations have to comply with the customs procedures that are applicable between the origin and the destination.

Another dimension is to view the maritime operations from the perspective of transport chain. Under this paradigm, there is a composition and shipment of the cargo through the rail route. Then the vessel operations are carried out, which are followed by customs clearance. Then the road route is used to deliver the cargo to the destination.

The third dimension is that of physical flows. From this perspective, the origin route is that of rail yard. Then the cargo arrives at the port. Then there is a transshipment hub in the maritime operations. Finally, the distribution center is used to deliver the cargo to the exact location.

From these three perspectives, it is quite evident that various regulatory frameworks come into force in the shipment process from origin to destination. The compliance of these frameworks has been challenging even in the traditional work mode for the maritime institutes and organizations. When all the systems are automated by AI-based algorithms, it will become even more complex for ship owners and ship managers to ensure compliance of the maritime shipping operations with the local and global regulatory frameworks.

Another issue in the regulatory paradigm is that the current workforce needs to be trained in the regulatory frameworks as well. The AI vendors develop products and display their functionality and prototypes on their websites. These functionalities might appear good at the first look but the ship owners, ship managers, maritime institutes and organizations need to review if a given functionality also meets the regulatory requirements. Training the workforce on the technical tasks of AI execution is already a big responsibility for the business managers. The situation becomes further complex when they are also required to provide training for the regulatory compliance.

The advancements in AI have also attracted the maritime shipping organizations. However, the ship owners, ship managers, maritime institutes and organizations should also consider the limitations of the vessel systems that are currently being operated globally. Various valuable insights were given at a recent seminar at London International Shipping Week. The guest speakers mentioned that more than 80% of the large vessels in the marine system do not have the basic communication facilities available during the vessel voyage.[xcv] It means that they do not possess the capacity of sending an ordinary email. In these circumstances, how the ship owners, ship managers, maritime institutes and organizations can expect to jump to the AI technologies where interconnectivity and cloud connectivity is the backbone of the whole system. The legal and regulatory barriers might outweigh the benefits that are to be received through cost savings.

IMO is currently evaluating various policies and procedures through which the operations of autonomous vessels could be regulated. However, at the same time, European Union is also formulating such regulations on an urgent basis.[xcvi] IMO has a track record of conducting deliberations at an extremely slow pace. As a result, there will be two regulations for the autonomous ships that might result in conflicting clauses at the national, regional, and international level. The adoption rate of AI-based systems in the maritime shipping industry is very low at present. Therefore, the regulatory frameworks may become a significant barrier to the wider expansion of AI-based systems.

The regulatory concerns may get converted to liability concerns for ship owners and ship managers when they implement AI-based systems. In the maritime shipping industry, when the AI-based systems are used by many maritime institutes and organizations, the courts might have to hear the cases regarding the ownership of data that is being provided as training dataset to the AI systems. Moreover, if the safety systems and predictions of an AI-based system fail, who should be held responsible for the losses incurred? These are some of the regulatory challenges that the maritime shipping industry will soon be facing when the implementation of AI-based solutions will gain momentum.

3.3. CULTURAL CHALLENGES FOR THE MARITIME INDUSTRY

AI implementation also poses cultural challenges for the maritime shipping organizations. It requires a significant business process reengineering and therefore, there should be a paradigm shift in the current organizational culture. The biggest challenge that the ship owners, ship managers, maritime institutes and organizations face is the resistance to change and a traditional mindset. The business processes of the maritime shipping are highly complex compared to the road shipment and air shipment. The industry has been relying on the traditional approaches for decades. Therefore, the crew members and other staff might show resistance to implementing AI-based technologies.

It can also be a challenge for the IT team to convince the senior management regarding AI implementation. The senior management will be of the view, 'do not fix it until it is broken'. However, it is the responsibility of the IT manager to make the senior management realize that many maritime shipping organizations have already embraced AI tools and technologies. If our organization does not opt for it, we will be simply out of the market. It will not take a decade, but we will be out from the market within years. The AI adoption is taking place at an accelerating pace, and carrying on with a 'wait and see' policy will not be prudent.

Another cultural challenge is to meet the training needs and skills gap of the current workforce. AI is all about the use of advanced, state-of-the-art technology and the end users should also have a good level of technological literacy. The skills gap of the current workforce should be filled regarding the understanding of the AI concepts. The training programs should also be arranged for up-skilling the current workforce. The organizational culture should be ready for embracing AI with its true potential. There is one more cultural challenge associated with AI implementation and it is the fear of losing the job. When the systems are automated, the current workforce will definitely have a realization that their services have become redundant. Think about autonomous ships, automated response systems, automated predictive maintenance, robots for container handling, robots for cargo handling, the fear factor in these circumstances is only natural. The ship owners, ship managers, maritime institutes and organizations should address these concerns of the current workforce. The employees should be made to realize that their services will be allocated to more strategic tasks and there will not be any retrenchment in the organization. Moreover, the implementation of AI will also improve their skill set regarding AI-based systems that reflect well on their CV. It is a highly demanded skill in the current maritime shipping context that will be valuable in the current organization as well as other organizations where the employee might work. AI tools and technologies only complement the human efforts and they have not been deployed in the maritime shipping organization to replace human crew members.

Another cultural challenge that might be faced is the collaborative work environment that becomes a necessity in the AI-based systems. The AI algorithms capitalize on the information collected from multiple sources connected through the internet. The receiving and availability of a high-quality data is paramount to the effective decision-making of AI algorithms. This availability of real-time, quality data can only be ensured if the maritime shipping staff works in a collaborative environment and all datasets are available in the integrated AI-based systems. This collaboration might become challenging in those work environments where the bureaucratic controls have been implemented and the managers take a lot of time in giving approvals and processing the information.

Another cultural challenge that might be faced in many maritime shipping organizations is the promotion of decision-making powered by data. The ship owners and ship managers are accustomed to intuition-based decision-making due to which AI-based implementations might sound a cultural shift in those organizations. Those organizations will have to promote a culture of intelligent decision-making and rely on data and AI utilization for an informed decision-making.

Another cultural challenge that might be faced is that the organizations are often influenced by short-term profit motives. AI implementation is a long-term investment for the maritime shipping organizations. If the business leaders of the organization do not have a visionary approach, they will find it extremely difficult to follow the whole process of AI implementation and AI integration. Therefore, AI implementation will also require the business leaders to change their leadership styles and become transformational leaders in the AI implementation process.

International Forwarding Association (IFA) has also mentioned various cultural challenges due to which the AI adoption rate is slower in the maritime shipping industry. According to IFA, the freight forwarders face the cultural barriers because the mode of operations has been different in different regions such as between the UK and the EU member states. In the case of UK, the emphasis of the government is more on stimulating innovation. In this regard, six principles have been mentioned by the regulators so that the maritime shipping operations could pass the criteria of transparency, fairness, and safety. There is much more flexibility in the UK regulatory framework to align the laws with the specific requirement of a given industry. However, contrary to this approach, EU has formulated an act known as EU AI Act.[xcvii]

This act does not believe in using a light-weight approach for AI implementation. The EU laws mention specific requirements for the users and developers of AI systems. An area of concern in the EU act for the managers and owners is that the system also allows the opportunity of claiming the compensation if the AI systems do not perform as expected and cause damage at the industrial or commercial level. It is a very heavy penalty from the perspective of maritime institutes and organizations because as I explained to you, in most of the cases, the users of the AI systems are not even aware how a particular recommendation or suggestion is being given by the AI systems. The understanding of the AI algorithms is a highly complex task, and the ship owners and ship managers are not expected to be proficient and tech-savvy to an extent that they could decipher the working of the AI-based systems.

Another challenge highlighted by IFA is the shortage of AI talent. The development of AI-based systems is a highly complex task and the AI programmers are available in a limited number at the global level.[xcviii] Those who possess the skills demand for a very high remuneration that adds a huge burden to the staff cost. At the same time, the senior management and the middle management of the maritime shipping companies might be good at the operations of shipment, but their knowledge of AI tools and technologies is very limited. As a result, the ship owners and ship managers fail to introduce AI in their specific organizational scenarios and contexts. They also find it challenging to make reliable and ethical decisions in the AI paradigm. The solution to these issues is more and more training of the ship owners, ship managers, and other leaders. As I keep on emphasizing to you, you cannot close your eyes to this AI-based revolution. Otherwise, your business will no longer be sustainable. The only way to gain a competitive advantage is to gain more and more understanding of how you can make AI work in your context and reduce the cultural challenges associated with the AI implementation.

As I explained to you above, the cultural challenges may emerge from multiple areas during the AI implementation in maritime shipping companies. These challenges can be addressed by an effective communication, transformational leadership style, and AI-based training and development of the workforce. The organization will need to promote a culture that welcomes AI implementation and optimizes the business processes through technological innovations.

4.   AI IMPLEMENTATION PLAN

The AI implementation plan for maritime shipping companies cannot be described in a precise series of steps such that the plan could be applied to all companies and organizational contexts. Therefore, in this chapter, I have mentioned several approaches that have been recommended by various organizations. You can read and review all these approaches and then you may prefer one implementation steps over other series of steps because it might be ideally suited in your business scenario.

4.1. KEY STEPS FOR INTEGRATING AI

Error! Reference source not found. explains one strategy where the key steps of AI implementation in the maritime sector have been explained by a data flow diagram. These series of steps have been recommended by IEEE in its newsletter. The figure highlights that the AI implementation should be seen from the perspective of the integration of big data and AI algorithms. The ship owners, ship managers, maritime institutes and organizations should focus on four key domains for implementing AI and optimizing the business processes.

The first process is digital transformation. As I explained to you in the previous sections, the vessel operations in the maritime region are still relying on old machines where it is even difficult to send an ordinary email during the vessel voyage. These systems should be transformed digitally so that the vessel systems are prepared for AI implementation. The digitalization should first be introduced in the maritime transportation so that the systems could communicate within different equipment and parts and to the external systems and interfaces such as internet cloud connections and satellite connections. The second area of the digital optimization is the systems deployed at different ports. If a maritime shipping organization is well organized at the origin side, it should not expect the same level of a conducive environment at the destination port. Digitalization will be needed also at the destination port for effective communication and data transfer through AI-based systems. The maritime transport system should not only be optimized from the current state but new innovations and systems should also be integrated into the current systems.

The second core area of intervention is to integrate big data into AI-based systems. As I highlighted in the previous sections, the effectiveness of the AI systems is highly reliant on the quality of data that is fed to them. This data should be big enough so that the system could learn most of the combinations and possibilities in the maritime shipping operations. The big data can be utilized for sustainability operations as well as maritime surveillance operations. In this context, IEEE recommends the use of support vector machine-based algorithms because these algorithms are highly efficient, effective, accurate, and precise in processing big data and make the systems learn from the big data.

The third core area of intervention is energy optimization and energy efficiency. The speed of the vessel operations should be optimized by the AI systems. The route optimization should also be enabled by the AI algorithms. As it is a well-known fact that maritime shipping operations are not limited to the vessel voyage between point A and point B. A key aspect of maritime shipping business is the effective handling of cargo and crane planning so that the cargo could be delivered easily and quickly to the place of the customer. Therefore, the AI algorithms should not only focus on route optimization at the maritime area but crane planning should also be enabled by the AI algorithms. In this regard, IEEE recommends that the business leaders should consider the use of particle swarm optimization algorithms because they can be more effective than support vector machine in this scenario.

The fourth core area of intervention is predictive analytics. The analytics system should first be used for viewing the performance of the vessel. The support systems should act proactively to correct the vessel operations before they reach the failed state. The second area of consideration is the implementation of a surveillance system such that all aspects of the vessel voyage could be seen visually through various dashboard indicators. The third area of consideration is to integrate the AI systems with the corporate applications so that they could also benefit from the results, recommendations, and suggestions of AI-based systems. In this area, IEEE recommends that the maritime shipping companies should use convolutional neural networks because they have a proven effectiveness in achieving a good level of predictive analytics. The results of the convolutional neural network and particle swarm optimization can also be used as key inputs by support vector machine algorithms. It also enables cross-cluster communality in the interfacing of different AI-based systems.

The model recommended by IEEE and mentioned in **Error! Reference source not found.** can become a good starting point for AI implementation in maritime shipping. These steps and processes should be considered as a guiding framework by ship owners, ship managers, maritime institutes and organizations for transforming maritime shipping for the future.

FIGURE 56: AI IMPLEMENTATION AND PROCESS BY MDPIxcix

MDPI, in a journal article, has also presented a roadmap for an effective AI implementation in four key steps. As shown in Figure 56 above, the first step is to gain a good level of understanding of AI tools and technologies and then assess the level of implementation needed in a maritime shipping company based on the organizational capabilities. The figure also highlights that ship owners, ship managers, maritime institutes and organizations will experience two key challenges in Step 1.

The first key challenge in step 1 is the presence of analog processes. AI communication and interconnectivity will require full digitalization of all processes so that datasets could be developed for the learning of AI algorithms. If a given maritime shipping company is still relying on analog processes, and these analog processes are big in numbers, then the AI implementation can only be carried out at an evolutionary pace. The second challenge that the ship owners and ship managers will experience is the level of transparency that they may expect in the AI implementation. As I explained to you earlier, the business managers will have a very limited understanding of how the AI algorithms are reaching at a particular conclusion. There might also be algorithmic biases present in the AI algorithms. Therefore, achieving a good level of transparency should be focused by the ship owners and ship managers and they might need to take technical advice from the technology professionals.

The second step is to review the current business model (BM) of the organization and identify the areas where the AI systems should be implementation and new business model implementation (BMI) is required. As I highlighted in chapter 3, AI has significance in multiple areas of maritime business. These may include logistics, supply chain, decision-making, safety, crew management, customer service, and preventing from cyber-attacks. I also explained to you that you cannot implement AI in all these systems in a single go. You will have to highlight those areas where the implementation should be made first and on priority basis.

The challenge that might be faced in step 2 is the misunderstanding regarding the potential of AI in a given area of maritime shipping. When the vendors are asked to explain their solutions, they will only highlight the strengths of their systems and how their systems are better than the other systems. Therefore, ship owners, ship managers, maritime institutes and organizations should contact multiple AI vendors. During their presentations, all stakeholders should be present and they should take notes regarding the usability of the system in the current organizational setup. Then, based on the input of all stakeholders, one of the systems should be approved by ship owners and ship managers.

The third step is to make the current systems capable and ready for the AI migration. For example, if the data regarding the health and quality of the vessel equipment is currently not being maintained, these datasets should be developed because they are going to provide key inputs to the AI algorithms. Similarly, if container and shipment details are not available in a digital format, they should be prepared so that a seamless integration and connectivity with AI-based systems is possible.

A real challenge in this step is also the level of transparency that may be achieved by ship owners and ship managers. For example, the ship managers might enable feeding a high volume of data to the AI systems without realizing that they do not have the consent available for using this data. The data-driven decision-making should not mean that the data should be used by the AI algorithms without the consent of the owners of the data. For example, if the personal information of the client is shared to the AI algorithms for an effective decision-making and business optimization, the ship owners and ship managers should also obtain an informed consent from the customers that their data will be used for AI analysis for improving our services.

The fourth and final step before the AI implementation is to gain an acceptance level of the AI implementation in the maritime shipping organization. As I highlighted earlier in the cultural challenges of AI implementation, the employees will view the AI initiatives with a sense of insecurity. They tend to believe that if everything gets automated, their presence will be redundant. The ship owners and ship managers should make them realize that their jobs will be retained and the AI systems are being implemented only for business optimization and business sustainability. The internal competencies also need to be developed because AI systems will be an altogether new interface for the maritime shipping professionals. They will need an extensive training for the proper usage and utilization of AI-based systems.

Several challenges may also be experienced by ship owners, ship managers, maritime institutes and organizations in step 4 of AI implementation. The first challenge is the lack of awareness of AI among maritime shipping professionals due to which their learning curve might be high and they might take an extended time for learning AI tools and technologies. The second challenge could be the lack of trust on AI tools because the recommendations will not be understood by the employees in the sense that they will not know how the AI algorithm reached at a particular solution. In the conventional mode, they are relying on the legacy systems and their own knowledge for making decisions. When their decisions based on intuition conflict with the data-driven AI solutions, they might be reluctant to implement the AI-based solutions. Due to this factor, I highlighted earlier that AI-based systems might show you only recommendations and suggestions and they might also be trained to implement those suggestions. The ship owners, ship managers, maritime institutes and organizations might go for a higher level of implementation if the employees do not benefit from AI-based solutions and still insist on making decisions based on their own judgment and intuition. When the employees see the benefits of following recommendations of AI-based systems, they will also accept the significance of these systems.

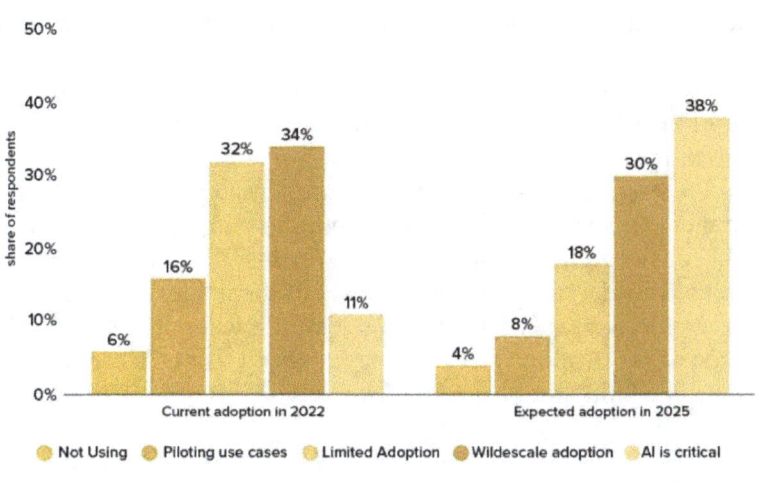

AI Adoption Rate in Supply Chain Globally: 2022- 2025

Figure 57: AI Adoption Analysis by AppinVentivc

Figure 57 shows a graphical analysis in which the current adoption rate of AI-based systems has been depicted by Appinventiv in global supply chains. The wide-scale adoption of AI was reported at 34% in 2022. The companies that felt that AI is critical and extremely important for their business sustainability were only 11%. Based on this current state of implementation, Appinventive has predicted the adoption rate of AI systems in 2025. Based on the prediction, the wide-scale adoption will be reduced from 34% to 30% in 2025. However, there will be a sharp increase in the count of those companies that consider AI as a critical success factor for their businesses. The percentage in this case will increase from 11% to 38%. This graphical analysis makes it evident that the implementation plan developed by ship owners and ship managers should consider the AI implementation as a critical factor and not just as a nice-to-have feature in the maritime shipping business.

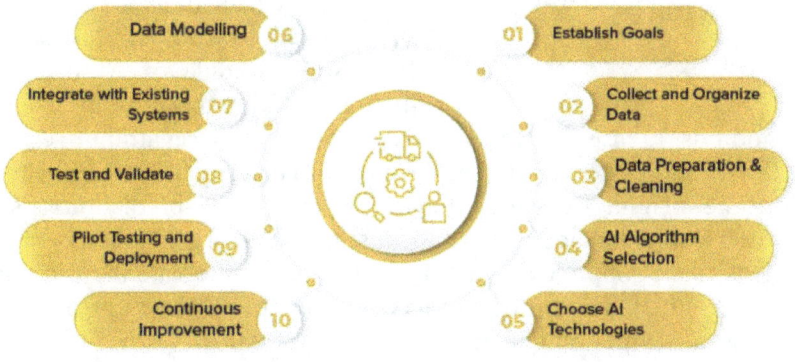

Steps to Optimize AI and Data Analytics in the Supply Chain

Figure 58: AI Implementation Steps by Apinventivci

Figure 58 highlights that Appinventiv is also one of these leading organizations that have recommended steps for an effective AI implementation and integrating AI into maritime businesses. Its recommendation is based on 10 steps. First, the ship owners, ship managers, maritime institutes and organizations should establish goals and objectives for AI-based implementation. They should identify areas of optimization such as route optimization, demand forecasting, and risk management.

The second step is the collection and organization of data. As I explained to you, the AI algorithms will need large datasets for an effective training. The relevant information should be collected and provided by the maritime shipping organizations from various sources. The pieces of information might include the records of the customers, the sales transactions based on customers and other categories, logistical data, inventory information, weather patterns, and market trends in the areas of operation.

The third step is to prepare the data and also clean it. You might be the first maritime shipping company in your area that is implementing AI. So, other systems are not familiar with the formats and standards of AI systems. Therefore you might also need to clean and prepare the data for the AI-based systems. The data should be checked for inconsistencies, missing values, and errors. The duplicate entries should be removed, and the data should also be properly formatted.

The fourth step is the selection of a particular AI algorithm. As I explained to you earlier in **Error! Reference source not found.**, different AI algorithms are used in constructing AI-based systems. You will have to select an algorithm that is particularly suited to your organizational context. The algorithms are broadly categorized into clustering algorithms, regression algorithms, classification algorithms, and deep learning algorithms.[cii] You might have to take advice from an AI expert. Explain to them the optimization requirements in your maritime business, and the expert will tell you which AI algorithm is best suited for you. The algorithm should have a high level of accuracy, precision, and it should be free from bias on the part of the AI programmer.

The fifth step is the selection of AI-based technologies. The AI technologies are available in different variants and you will have to select from them for the maritime shipping business. In the intensive level of AI implementation, the company might consider autonomous ships and robots for container handling. However, other technologies should also be considered in the initial phases of the implementation such as computer vision, robotic automation, predictive analytics, machine learning technologies, and the technologies offering natural language processing. Appinventiv has recommended five core areas where the AI should be implemented in the maritime shipping business. These core areas are highlighted in Figure 59 below:

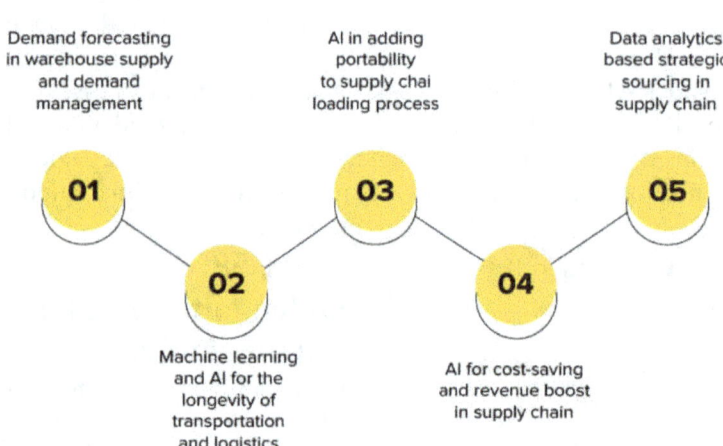

FIGURE 59: FIVE CORE AREAS OF AI INTERVENTION[ciii]

The first major area of intervention is demand forecasting. The maritime shipping companies may allocate more resources in their business operations during the off seasons. Their business operations should always work with optimal resources whether it is seasonal sales or off season. The demand forecasting module of AI can be highly helpful in this regard.

The second major area of intervention is to use AI for the recommendations regarding vessel maintenance. For this purpose, an IoT-based AI system should be developed. There should be a delivery of real-time data and transfer of real-time insights from the AI systems. The failure predictions also provide valuable assistance to ship owners, ship managers, maritime institutes and organizations.

The third major area of intervention is to make supply chain highly portable and automate the loading process. The AI systems not only automate the loading process but also provide real-time visibility to the loading process. If, in the initial process, the robotic implementation is not feasible for the maritime shipping companies, then they may at least get real-time visibility to expedite the process protocols and manage the parcels with a lesser level of risk.

The fourth area of intervention is to use AI for boosting the revenue and profitability. You should make all your business decisions based on intelligent data and reports so that the maximum revenue could be earned from the cargo orders and contracts. AI has a proven track record of providing these revenue boosts. In a Bloomberg Report, it has been acknowledged that the cost of moving goods by the maritime route has increased by 12% in the recent years as shown in Figure 60 below. The use of AI systems can make a significant reduction in the cost of shipment.

FIGURE 60: INCREASE IN THE COST OF GOODS SHIPMENT BY SEA[civ]

The fifth area of intervention is the use of data analytics. The conventional data analysis systems offer a lower level of functionality because they do not possess the capability of processing big data and they

Choppy Waters

The cost of moving goods by ship has soared to a 5 1/2-year high

/ World Container Index container freight benchmark rate

$2,200 per 40-foot box

Source: Drewry World Container Index

Bloomberg

do not have a mechanism of training and learning from the datasets. Therefore, AI-based data analytics systems should be used by ship owners, ship managers, maritime institutes and organizations for a strategic sourcing, strategic partnership, and supplier relationship management. The performance indicators can also be reviewed for compliance in real-time. The system also suggests lower-cost alternatives for the shipment process as well as route optimization.

The sixth step is known as data modelling. As I explained to you earlier, the training is considered as a pre-processing part of AI algorithms. When the algorithms are properly trained, they build a model for decision-making, and it is in fact this data model that interacts with the users of the AI system. In the maritime AI contexts, some of the state-of-the-art prediction models have been developed that you can also utilize in your organization. For example, two prominent algorithms are auto-encoders and seq-seq that produce forecasts with a high level of accuracy.[cv]

The seventh step is to integrate AI-based systems with the existing systems of the maritime shipping company. If the existing systems follow the global standards and the best practices, they can easily be integrated with the AI systems. Therefore, I would recommend ERP (Enterprise Resource Planning) systems in your maritime shipping organization for a better integration. You can acquire various key modules such as Transport Management System (TMS) and Warehouse Management System (WMS). ERP systems are offered by the leading vendors such as SAP, Oracle, and Epicor.

The eighth step is to test and validate the functionality of AI-based systems after implementation. The testing is a very crucial step because the AI-based systems may go wrong if the data quality is low and the proper training is not done before developing the data model. As I mentioned earlier, you will be challenged in knowing the dynamics of how the AI system arrived at a particular solution. However, in the testing phase, you can evaluate how the recommendations varied from the ideas of the experts and whether the recommendations pass the criteria of dependability and precision. The QA experts should play a pivotal role in this phase and they should emphasize on iterating and improving the models based on the concerns raised by the end users.

The ninth step is to conduct a pilot testing of a system before making the system fully functional in the live environment. The pilot testing is conducted on a small subset of the real systems and business processes. The issues highlighted in the pilot testing should also be rectified by the AI experts.

The tenth, and the last step, is to focus on the continuous improvement of the AI-based systems. As I mentioned earlier, the data model of the AI system keeps improving as it encounters with new data sets and the live data. You should capitalize on this opportunity and make continuous improvements in the implementation of AI-based systems.

Deep Sense AI has recommended six steps for building and executive an effective data model as shown in Figure 61 below. As I highlighted earlier, the data model is the core of the AI-based systems. The learning of the data model determines the effectiveness of the AI algorithms. Therefore, these six steps should be implemented to develop an effective data model. Once the data model is able to execute a fair level of decision-making, it should be put into operations. Over time, through learning and interactions with the user data, the model will fine-tune itself and the level of recommendations and suggestions will be optimized.

Figure 61: Six Steps of Implementing AI Data Model

The first step is to develop a good level of understanding of the business. The maritime shipping business professionals already possess a good level of the understanding of their business domains, but they will need to view the business from the perspective of AI-based systems. They will have to translate their business requirements into the systems and interfaces that they need in the AI-based systems. They

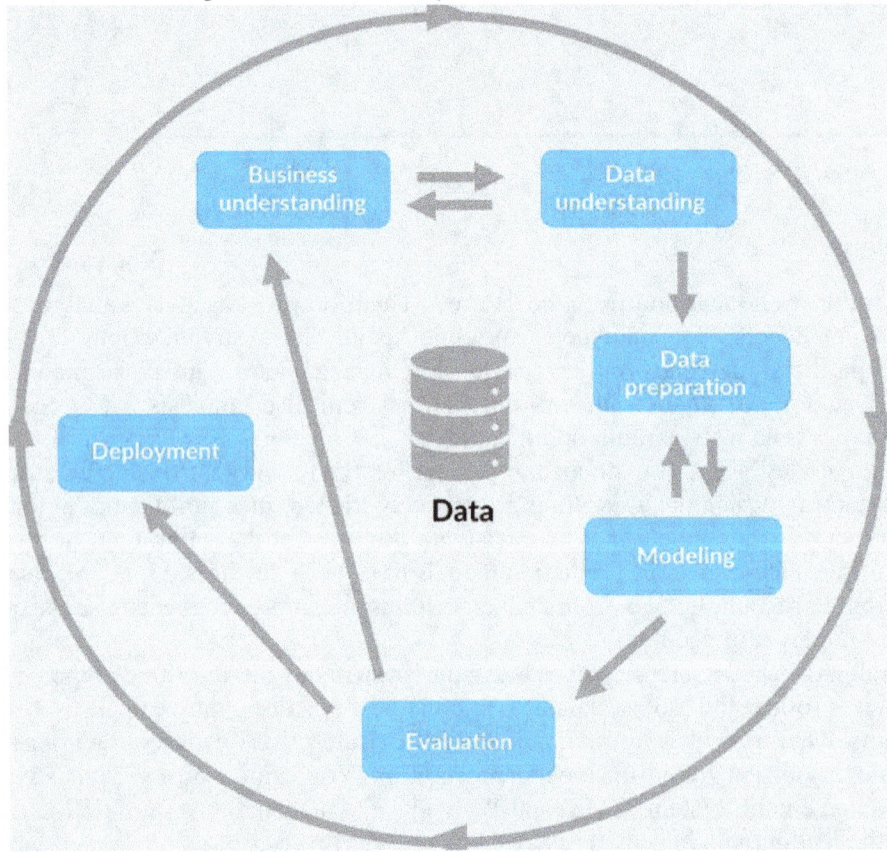

may consult the case studies and success stories and demand from the AI programmers that the same level of functionality is also needed in their maritime institutes and organizations.

The second phase is the understanding of the data. As I keep emphasizing throughout this book, the performance of the AI algorithms is all about the quality of data it receives. The ship owners and ship managers should focus on the readiness of the data for the system migration and also ensure the full transparency concerning the acquisition of data that may also include the receiving of an informed consent from the relevant parties.

The third step is the preparation of the data. At the end of the day, the AI-based systems also save the data in the databases such as oracle and sql server. The tables and fields are defined in those database platforms. The data from the existing systems should be prepared in a format that is consistent with the new systems and can be easily imported into the new AI-based systems.

The fourth and the most important phase is data modeling. Deep Sense suggests two approaches for building the data model. In the first case, the ship owners and ship managers can use a proven and tested approach for data modeling. In the second case, the maritime institutes and organizations may explore several data models in parallel and select the one that provides the best accuracy and precision in the recommendations and suggestions.

The fifth phase is the evaluation of different use cases that have been developed by the AI programmers. The execution of the use cases will make the ship owners and ship managers a real feel related to the significance of using AI-based systems. They will see in real-life scenarios how the cost savings are being accomplished and how the business processes are optimized.

The sixth, and the final phase, is the deployment of the AI-based solution. The deployment should be made after an extensive testing of the system. The system should be tested based on the unit testing, the whole system testing, as well as the integration testing. If all these phases suggested by Deep Sense are executed efficiently, then a good AI-based infrastructure will be in place that will transform maritime shipping for the future.

GlobeMA emphasizes on using a few steps in the AI implementation plan as shown in Figure 62 below. The good aspect of the strategy recommended by GlobeMA is that alternative paths have been mentioned at each step. So, for example, if plan A fails in any AI implementation, the ship owners and ship managers can go for plan B.

FIGURE 62: AI IMPLEMENTATION STEPS BY GLOBEMAcvi

The first step in this approach is the development of a business hypothesis. For the development of hypothesis, first the interviews are conducted with the customers, employees, suppliers, and other stakeholders. The need for the AI-based implementation might be felt by the ship owners, ship managers, maritime institutes and organizations for various factors, and all these factors should be converted into hypotheses. As mentioned in the figure, the first hypothesis could be that the customers are not satisfied with the quality of data available to them on the physical and online places to know different service offerings of a maritime shipping organization. The second hypothesis could be that the existing solutions such as logistics, crew management, cargo handling are not producing satisfying results. The third hypothesis could be that the customers have a high expectation with a maritime shipping company that the company will soon incorporate AI in its business operations.

The second step is the conversion of the business hypothesis to a business case. If a maritime shipping company feels that the benefits of AI implementation will indeed be materialized, but the cost is too high, and their current budget does not allow for allocating so much financial resources, then the project might be stopped by the AI developers. However, there is also an alternative route in this case, where the AI developers may recommend a partial implementation of the AI solutions in the areas of critical need.

The project, however, will stop and there will be no alternate plan if the business hypothesis is not validated by the senior management and the management does not see any tangible outcomes of the AI implementation. If there is a positive verification from the management side and the budget is also available for the AI implementation, then it will be considered as the best-case scenario by the AI developers. In such cases, the next step, step 3, will be the full implementation of the AI-based solutions. If the management only allows for partial implementation, then the partial modules will be implemented in those maritime shipping companies. Those modules will be already agreed upon modules and further implementation will be subject to the availability of funds.

This approach by GlobeMA is quite impressive in that it offers a lot of flexibility in the implementation plan. The business hypotheses are developed by the IT and functional team after a lot of deliberations. It is then highly likely that these hypotheses will get the go-ahead of the senior management. The only area where the management shows reluctance is the availability of funds. However, this framework even caters this limitation by enabling the partial implementation of the AI-based modules in the maritime shipping companies.

In the above implementation steps recommended by renowned institutions and organizations, you will have got a clear sense of direction that as a ship owner, ship manager, maritime institute or organization, how you can implement and integrate AI in maritime shipping in the best possible way. You will also have realized that such an implementation in maritime shipping will result in significant improvements in safety, efficiency, and competitiveness of the maritime shipping company. In your specific organizational context, a well-executed and a structured plan will be highly crucial for a successful AI integration. You should consider several factors for ensuring success and effectiveness in AI implementation.

The first area of consideration is that you should define clear objectives for your AI implementation. Which improvements and efficiencies you are expecting to achieve in maritime shipping through AI implementation? This is the question that the business managers will need to ask again and again. You should consider several areas for optimization such as improving the efficiency of the crew members, enhancing the safety of the vessel operations, and optimizing the efficiency of the fuel consumption. When your objectives are clear and focused, these objectives will then provide guiding principles for the entire project.

The second area of consideration is the assessment of the IT infrastructure and data sets availability. You should evaluate your current IT infrastructure, systems, and data sources as to whether they are ready for AI implementation. The data sources should not only be evaluated regarding availability but also concerning the quantity, quality, and accessibility of the data. Your IT infrastructure should be robust and resilient to support the deployment of AI-based systems and AI upgrades.

The third area of consideration is the preparation and collection and data. The collection of data will not finish the task of maritime shipping professionals. The data should be preprocessed and cleaned to get it ready for feeding into AI algorithms. The power of data cleaning should not be underestimated by ship owners, ship managers, maritime institutes and organizations. There is a huge reliance of the AI algorithms on the high-quality data, therefore, the business managers should prepare the datasets accordingly. They may have to structure the data as well as doing the labelling of data so that it can easily be used by the AI algorithms for the training purposes.

Another area of consideration is the selection of suitable AI technologies. The definition of 'suitable' will vary from one maritime shipping organization to another, and therefore, you will need to do an extensive analysis at this stage. In the first step, you had set clear objectives for your AI implementation. Now you should select those AI algorithms and AI technologies that could best meet your set objectives. You will have to review and assess among different subfields of AI. These may include deep learning algorithms, machine learning technologies, computer vision, natural language processing technologies, and other related domains.

Once you have selected AI tools and AI algorithms, you should move on towards the proof of concept and pilot projects. In the initial phase, consider the use of small projects and test the acquired AI applications on these projects by using a controlled environment. The use of small projects will assist you in evaluating the feasibility of the AI applications for the maritime shipping organization. You should also evaluate how the acquired AI application is making an impact on the business processes of the maritime shipping organization and which type of modifications and adjustments will be needed when the project will be deployed at a full scale.

Another area of consideration that might improve your chances of an effective and a successful AI implementation is the level of collaboration that you make with the AI experts. You should build collaborations with the data scientists, AI experts, and the development teams of AI. You will also have to make a strategic selection regarding the training of AI applications. One option could be to outsource the AI experts for conducting in-house training sessions for the managers and the crew members. The other option could be to develop master trainers from within the company and ask them to train the entire crew. Your decision of selecting any of these two options will be heavily reliant on the organizational skill set and the availability of resources.

Another factor that you will have to consider as ship owners, ship managers, maritime institutes and organizations is the regulatory compliance of the AI implementation. The implementation should be compliant with the local and international maritime regulations. You should also evaluate the regulatory compliance of the datasets and the implementation should be compliant with the local and international data privacy laws. You might have to get the advice of a legal counsel in this regard. The lawyer will review the entire regulatory landscape and highlight the issues that may emerge in the context of data protection and liabilities.

Another crucial aspect in the AI implementation is the training and skills enhancement of the workforce. The training programs should be arranged to ensure that all crew members have attained a good level of AI literacy, and they are in a position to operate the AI applications efficiently. The workforce should be provided training not only on using AI tools but also on using the reporting tools and AI-based insights.

One area of concern that should always be evaluated by ship owners, ship managers, maritime institutes and organizations is the issue of data privacy and data security. When you expose a huge dataset to gain deep insights, protecting the privacy and confidentiality of the data is also your responsibility. You cannot just enter into the AI domain with a positive mindset that any privacy and data security issues will be handled as and when they emerge. You should be highly proactive in this regard and introduce robust cybersecurity measures. These initiatives should be capable of protecting the sensitive maritime shipping data that is being utilized by the AI applications. You should also review all privacy laws and regulations that apply to the vessel operations and shipping and ensure strict compliance to reduce the risk of fines and penalties.

A critical factor in AI-based implementations is that the performance of the systems evolves over time when they encounter new data and learn from the interactions. Therefore, the performance of the AI systems should be continuously monitored by ship owners, ship managers, maritime institutes and organizations. You should develop a plan for a continued maintenance of the system and applying updates to access the latest versions of the AI-based system. There should also be retaining schedules arranged by the IT department so that the AI model can improve learning based on the new data points. It will facilitate in improving the relevance and accuracy of the AI model.

As I highlighted in chapter 4, cultural challenges may also be faced in the AI implementation in a maritime shipping company. Therefore, the cultural transition of the AI project should also be managed effectively by ship owners, ship managers, maritime institutes and organizations. The managers should make the crew members realize the significance of AI systems and they should also address the concerns of the crew in this regard. There should a conducive environment and an enabling culture in the maritime shipping organization where the implementation of AI-based systems is viewed as an enabler and a facilitator instead of viewing them as a threat to their jobs.

The AI implementation should also be improved by the introduction of KPIs and evaluation metrics. I have explained comprehensively in this book that AI algorithms process a very large dataset for reaching the conclusions, and in most of the cases, it will not be possible for ship owners, ship managers, maritime institutes and organizations to know the calculations behind the scene. Therefore, performance indicators should be developed by the ship owners and ship managers for tracking the progress of AI implementation. The performance should be periodically evaluated regarding the accuracy and precision of the results and the impact that the AI implementation made on transforming maritime shipping for the future.

Once you have achieved success in implementing AI in one of the business areas of maritime shipping, it is time to scale up AI integration. The scope of AI implementation should be expanded to other areas and business processes to gain a competitive advantage. During the expansion phase, the ship owners, ship managers, maritime institutes and organizations should also consider the lessons learned from the previous implementations so that more effectiveness could be achieved in the implementation phase.

Another area of consideration that you should consider during the implementation plan is the stakeholder communication. The port authorities, other vessel operators, data providers, and government agencies are the key stakeholders in the maritime shipping operations besides crew members and the customers. All the stakeholders should be updated regarding the AI-based transformation of the maritime shipping company. The project progress and the information regarding the upcoming implementations should be shared with all the relevant stakeholders.

The last but one of the most important steps in the implementation plan is gathering feedback from the stakeholders and the employees. They should be encouraged and motivated for providing feedback for the continuous improvement of the maritime shipping organization. The ship owners, ship managers, maritime institutes and organizations should use the received feedback for improving the application interfaces in AI-based systems. The systems should be evolved, modified, and optimized based on the feedback. There should also be adaptations and adjustments in the system based on the changing technologies and requirements of the organization.

The above discussion on AI implementation plan and the key steps for integrating AI makes it evident that a successful implementation is possible through dedication, commitment, and a careful planning. The ship owners and ship managers should develop a culture that welcomes AI-based innovation. By following the steps of the implementation plan, maritime shipping organizations can receive the full benefits of AI implementation. AI-based implementations should result in improved efficiency, safety, and competitiveness of the maritime shipping organizations.

4.2. RISK MITIGATION STRATEGIES CONCERNING CHALLENGES AND BARRIERS

In chapter 4, I have explained in detail various challenges and barriers to AI implementation. I presented the challenges in three key categories of technical, regulatory, and cultural challenges. However, I also emphasized that facing the challenges does not mean that ship owners, ship managers, maritime institutes and organizations should stop implementing AI in maritime shipping organizations. These challenges can be addressed effectively, and I also mentioned case studies, where the maritime shipping organizations were successful in their implementations despite facing these challenges. In this section, I am highlighting the risk mitigation strategies for eight key challenges and barriers that are often faced in AI-based implementations.

The first, and the topmost challenge, is the availability of the dataset, and the available of that dataset, which is of high quality and can enable a good learning process for the AI algorithms. The risk is that if the data is of low quality or inconsistent, the accuracy and precision of the AI data model will be low. This risk can be mitigated by using several risk mitigation strategies. The ship owners and ship managers should make a significant investment in the acquisition of the required dataset.

Figure 63: An Approach to Risk Mitigationcvii

Figure 63 shows an approach of risk mitigation in maritime shipping organization concerning AI implementation. The risk mitigation approach should be categorized into four phases. In Stage A, the ship owners, ship managers, and the maritime shipping staff should prepare themselves for the evaluation of the threats. The evaluation should be based on a systematic review of the whole system. Based on the review, the possible attack groups should be identified. In the next step, the vulnerabilities should be identified in various components of the vessel system and the shipping system.

In Stage B, the maritime shipping organizations should focus on the identification of scenarios. The possible attacks should be classified into different known and unknown attack types. The consequences should also be documented if the attack becomes a reality.

In Stage C, the ship owners and ship managers should rank different scenarios and possibilities of attack. They should estimate the likelihood of each attack and attack type. The consequences should also be ranked in the similar manner.

In the final stage D, the control barriers should be identified from the context of maritime shipping operations. The risk assessment should also consider the possibility of the emergence of new scenarios of attacks. The net outcome of all this activity of risk assessment should be the development of safety recommendations and these recommendations should be strictly enforced by ship owners, ship managers, maritime institutes and organizations.

FIGURE 64: CLASSIFICATION OF AI RISKScviii

Figure 64 above classifies the risks associated with AI-based system implementation and this classification has been presented by Harvard Business Review. This classification system indicates that the AI-based risks may be experienced at the levels of application systems and/or at the business and national levels.

Three key application level risks are performance risk, security risk, and control risk. Performance risks may be observed in AI-based systems when the system is unstable and the recommendations of the systems vary when similar events occur, for instance, two to three times. It may reduce the trust level of the users on the utility and effectiveness of the system. Another risk that I also mentioned earlier is the algorithmic bias that may be experienced in AI-based implementations. The system may also show errors and bugs because the code is written by a human AI developer and logical and syntax errors may be present in the AI programs. Therefore, the risk mitigation strategies should focus on reducing the performance level risks of the AI systems.

Another application level risk is known as the security risk. AI-based systems will be deployed in an interconnected, cloud-based environment. Therefore, privacy risks, cyber attack risks, and open source software risks have the highest likelihood. The controls should be built at the application level to mitigate these risks.

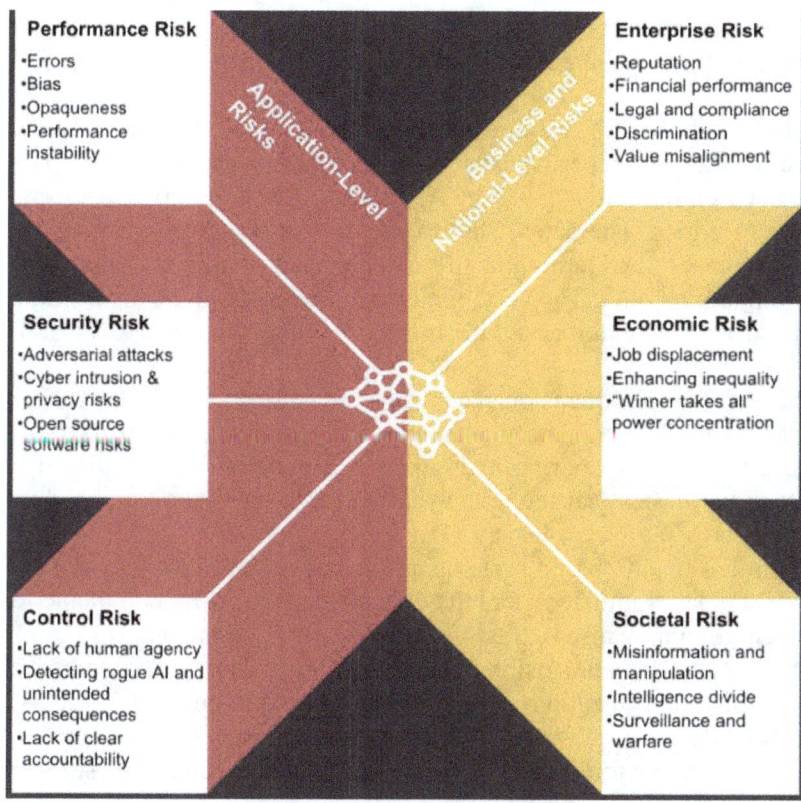

Performance Risk
- Errors
- Bias
- Opaqueness
- Performance instability

Security Risk
- Adversarial attacks
- Cyber intrusion & privacy risks
- Open source software risks

Control Risk
- Lack of human agency
- Detecting rogue AI and unintended consequences
- Lack of clear accountability

Enterprise Risk
- Reputation
- Financial performance
- Legal and compliance
- Discrimination
- Value misalignment

Economic Risk
- Job displacement
- Enhancing inequality
- "Winner takes all" power concentration

Societal Risk
- Misinformation and manipulation
- Intelligence divide
- Surveillance and warfare

The third application level risk is control risk. As I highlighted earlier, when the AI-based systems are efficient and optimized, all the stakeholders are happy with the AI implementation. However, if something goes wrong or if something unexpected happens, it becomes extremely difficult to identify the responsibilities and liabilities because multiple systems interact with one another in the AI-based environment. The misappropriation is also possible in these systems. There is also a concept of rogue AI where they system behaves in an undesirable manner. At that time, it becomes difficult for the handlers to minimize the consequences of rogue AI behavior. Therefore, control risk strategies should also be developed as part of the overall risk mitigation strategies.

The first business level risk is known as enterprise risk. If the systems do not demonstrate the desired efficiency, the goodwill and reputation of the maritime shipping organization may be at stake. The financial performance of the company may also be influenced negatively because a huge cost is incurred in developing the infrastructure for AI-based implementation. The organization may also face legal and compliance issues particularly when the data acquisition from multiple sources is involved. The discrimination may also be observed in the maritime shipping organization where the ship owners, ship managers, maritime institutes and organizations prefer AI chatbots and robots over the human efforts and human-based interventions.

The second business level risk is economic risk. AI-based tools and technologies may execute numerous tasks currently carried out by humans. Therefore, job opportunities may be reduced in the maritime shipping organizations due to an enhanced level of automation. The power dynamics in the organizational setup is also affected where the key business decisions are not based on the input of the senior management but on the recommendations of the AI algorithms.

The third business level risk is societal risk. Due to the use of large datasets and acquisition of data from multiple sources, there is a high possibility of manipulation and misinformation. As I explained earlier, the AI algorithms may be compelled to make judgment errors by feeding incorrect data for the learning of AI model. It poses serious threats particularly in cases when the AI-based systems are used for surveillance in the maritime regions.

FIGURE 65: DIMENSIONS OF AI RISKScix

Figure 65 highlights that risks may be experienced in AI-based implementations from multiple dimensions. These dimensions together create an identity risk for the maritime shipping organization. Therefore, the ship owners, ship managers, maritime institutes and organizations should mitigate the risks associated with all these dimensions for a successful AI implementation.

The first dimension is known as the strategy dimension. The fate of an AI project may also go wrong even at the strategy level. As I highlighted, various vendors, projects, and areas may be selected by ship owners and ship managers for an initial AI implementation. If a wrong project is selected, then all the AI investment may get wasted. If the organization does not focus on creating a sound digital infrastructure for the project, then it may also pose a risk. If the project managers do not coordinate with all stakeholders and the support of the senior management is also not available to them, then the project implementation is at risk.

The second dimension is called trust dimension. The functionality of an AI-based system resembles a black box where the end users have a little idea regarding the working of the algorithms. The user experience might be poor if they are not well trained in using the AI-based systems effectively.

The third dimension is the dimension of ethics. The dataset fed to the AI-based algorithms may not represent the whole population. In those cases, the learning of the AI algorithms will be wrong and it will also reflect in their decision-making approaches. The system may also suffer from the algorithmic bias and the data bias. Due to the higher level of automation, the human oversight is either reduced or missing in AI-based implementations.

The fourth dimension is known as the compliance dimension. The data acquisition may violate the data privacy laws of various countries and regions. The regulatory oversight may also get reduced due to the higher level of automation. The legal team in the maritime shipping organization may also find it challenging to assess the compliance of the laws because the lawyers will struggle in comprehending the dynamics of AI-based systems.

The fifth dimension is the financial dimension. AI-based systems and infrastructure development are highly expensive. If there are delays in the data acquisition and the development of AI model, then cost may increase even more. The implementation may also exceed the planned cost and budget.

The sixth dimension is known as the technology dimension. The technical skills of the IT and the functional team in the maritime shipping organization may not be up to the mark for the AI implementation. The error analysis may also be weak. In those cases, there will be a huge reliance of the maritime shipping organization on the AI vendor that will increase the maintenance and support cost.

The seventh and the last dimension is the people dimension. If there are extreme weather conditions, the AI algorithms might have to be reworked to calculate all parameters that will increase the workload of the crew members. The organization may also face resistance from environmental protection organizations for implementing technology-intensive projects in the maritime region.

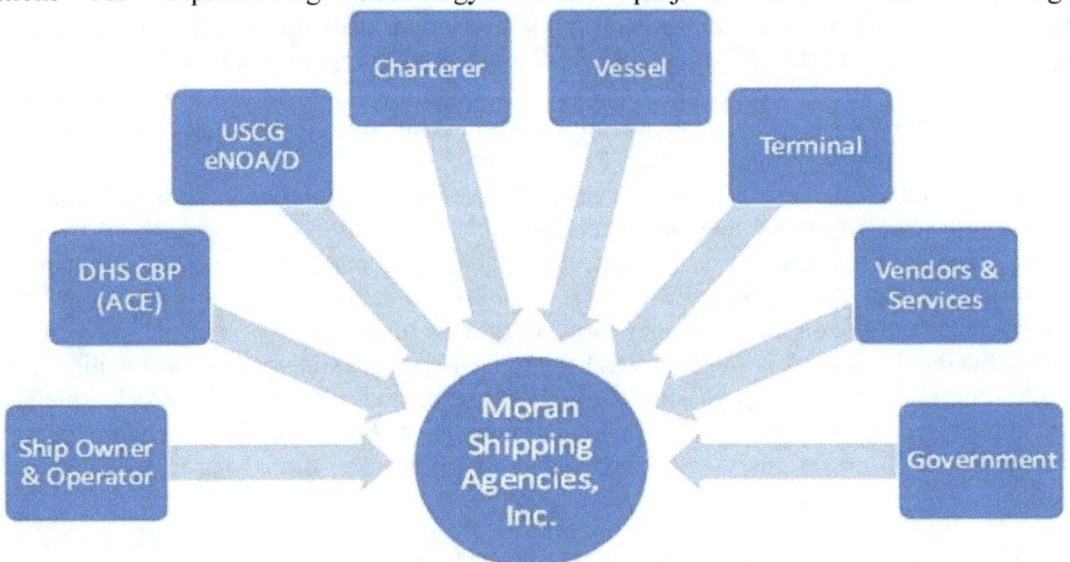

**Maritime Sector Has a Complex and Sensitive Ecosystem of Data
(Source: Moran Shipping)**

FIGURE 66: COMPLEX ECOSYSTEM OF MARITIME AIcx

Figure 66 shows that the maritime ecosystem is highly complex. When the touchpoints of the maritime shipping are connected by IoT-based and AI-based systems, they cybersecurity risks emerge regarding the integrity and security of data.

The ship owners and ship operators conduct vessel operations based on the best available information and the experience. The next interaction is with DHS or Department of Homeland Security. This interaction is also critical and the AI-based systems should ensure the compliance of the AI systems with the safety parameters of the country. The next interaction is with USCG or cost guards and they will also evaluate the compliance of the AI-based systems. Then there are charterers and the vessel itself that should be checked through predictive maintenance systems of AI. The vessels then arrive at a particular terminal and the regulations concerned with a particular terminal should also be complied with by the ship owners, ship managers, maritime institutes and organizations.

The AI-based system will also interact with various vendors and services for intelligent decision-making. The data from the government agencies may also be fetched by the system. These all touchpoints should be considered by the IT team of the maritime shipping company for preventing the system from cyber attacks.

FIGURE 67: RESPONSE OF AI FOR MITIGATING RISKScxi

Figure 68shows various strategies, protocols, and functionalities of AI-based systems in maritime shipping. The ship owners, ship managers, maritime institutes and organizations should ensure the incorporation of all these functionalities in the AI-based systems for mitigating the risks and providing an effective response to the emerging threats in the maritime region.

AI systems should also be equipped with the earth observations and data so that they could provide real-time hydrological analysis to the ship owners and ship managers. The system should also have the intelligence of an efficient disaster decision-making. The illegal activities and public health concerns should also be automatically reported by AI systems. The system should be supported with water-saving devices and there should be a decentralized mechanism of services. The system should also simulate the concept of urban digital twins so that there is an advanced water-sensitive analysis.

In order to reduce the risks in the maritime region, the system should have forecasting interfaces for neural ESM and climate risks. The ship operators should also possess mobile water testing kits. The system should work on the principles of sustainable water management. The system should have a built-in intelligence for blockage diagnosis and water pipe leak. The system should also enforce virtual forensics and testing. The pump station maintenance should be carried out by the predictive maintenance feature of AI. The performance of the treatment plan should be quickly optimized based on the available parameters.

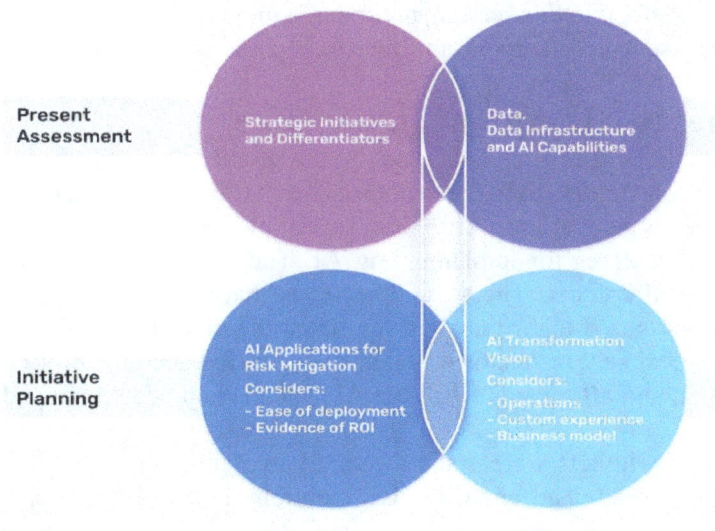

AI Initiative Prioritization in Crisis

Figure 68: Refinement of Risk Assessment Approachescxii

Figure 68 highlights the flaws in the current approaches to risk management and how these risk management approaches should be optimized for AI initiatives. In the conventional approaches, the strategic initiatives are formulated and the differentiated aspects of the AI systems are highlighted. Then the focus is shifted to AI data acquisition, data infrastructure development, and building AI capabilities. However, the recommended approach presented by Emerg is to focus on the evaluation of risks and develop risk mitigation strategies first. At that time, the ship owners, ship managers, maritime institutes and organizations should focus on the evidence of ROI and how there could be a seamless deployment of the AI solution. Then a transformation vision should be developed for AI-based implementation. Under this vision, the ship managers and ship owners should consider the operational requirements, the current business model, and the customer experience to be enriched by AI-based implementation.

Figure 69: Barrier-based Approach to Risk Managementcxiii

Figure 69 shows a barrier-based approach to risk management that is known as bowtie method. Under this approach, a top event is first identified as a hazard. Then preventing barriers are created to prevent the threat from influencing the top event. The escalation factor is also considered and an EF barrier is also created.

If the top event is still affected by the threat, then recovery barriers are created to minimize the impact of consequences. An EF barrier is also created to reduce the impact of the escalation factor. This approach is an excellent strategy for viewing the risk from both threat and consequences perspectives. The barriers are created from both perspectives so that the AI-based system is always responsive and resilient.

FIGURE 70: RISK MANAGEMENT FRAMEWORK BASED ON SUSTAINABILITY PRINCIPLE AND CLAUSEScxiv

Figure 70 illustrates the risk management framework presented by MDPI in a journal article. The framework is based on the sustainability clauses. The framework argues that the risk management strategies should be developed by first considering the principles of sustainability such as system approach of the AI system, addressing uncertainty, and ensuring transparency. Based on these principles, a risk management framework should be developed that should have a cyclic approach.

The framework should be based on the mandate provided by ship owners, ship managers, maritime institutes and organizations. The risk management approach should be based on the design of the framework. The framework should undergo continual improvements. A steering committee should review the approaches of the framework periodically.

The framework should be a key input to the processes of the risk management and the risk mitigation strategies. While developing the processes, the ship owners and ship managers should first establish the context. The risk assessment should include the processes of identification, analysis, and evaluation. Then risk mitigation should be accomplished by a continued process of communication, consultation, monitoring, and review.

4.3. SUCCESS STORIES OF AI IMPLEMENTATION

I have mentioned case studies of successful maritime shipping organizations that embraced AI and I have also described a list of renowned AI vendors. In this section, my focus is on highlighting the secrets of success when the AI implementations went very smooth and the organizations benefitted greatly from the AI implementations.

Figure 71: Optimal Approaches to AI Implementationcxv

Figure 71 highlights the optimal approaches that were adopted in the maritime sector and the organizations were highly successful in implementing AI. These approaches have been highlighted by Valuecoders in their analysis of various organizations. The first secret of success is that those organizations formulated clear business objectives for implementing AI. They were clear regarding their benefits specific to their organizational context. The second secret is that they focused on the acquisition of quality data and ensured that the AI algorithms are free from the algorithmic biases on the part of the AI developer. The third critical success factor was that they ensured accountability and transparency during the AI implementation and afterwards when the AI systems had been deployed at the production level.

Another success factor was that they started with a few areas and the small level implementation of AI and then scaled up the implementation process. The analysis also mentions that those companies invested heavily in talent management and infrastructure development. Those companies are also

78

characterized by the fact that monitored the performance of the AI systems on a continued basis and made adaptations in the system accordingly.

Figure 72: Reasons of Failures of AI Projectscxvi

Figure **72** shows the results of a survey conducted by Gartner. The findings of the survey revealed main reasons due to which the organizations fail to integrate AI in the company. The key reasons were the lack of skills, selection and execution, tools, productization, and governance.

The productization emerged as the major challenge because it was difficult for the organizations to deploy AI into their legacy business processes and applications. This barrier was highlighted by 47% of the participants in the survey. The participants also indicated that most of the AI-based systems are being developed by using open-source software and these software platforms are not capable to deliver production level results.

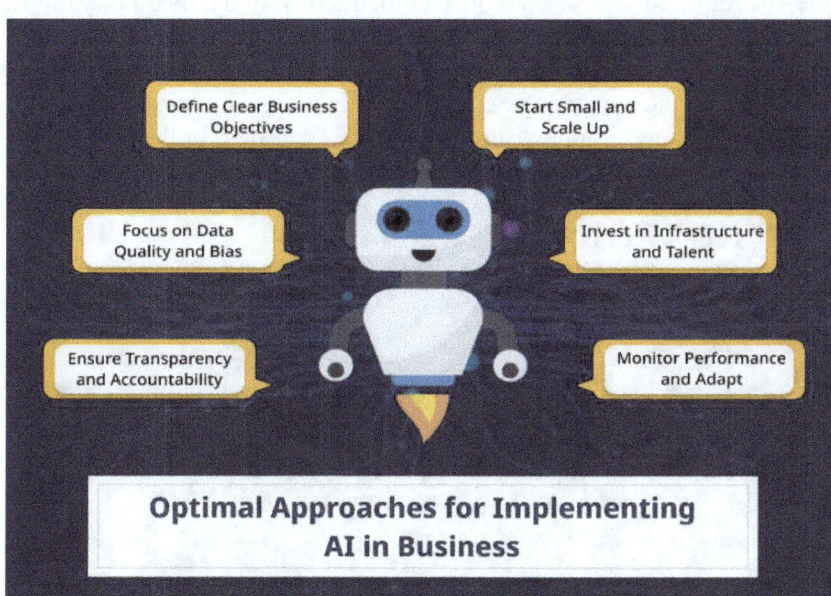

The second major challenge was selection and execution. During the implementation phase, 36% of the participants reported that they faced resistance from the employees. The management also had unreasonable expectations from the AI implementation and they were also unable to calculate an approximate ROI for AI implementation.

The third major challenge was concerning tools. A total of 20% of the participants reported that they did not have an idea of the right tool for the AI implementation or they do not have necessary funds for the procurement of those tools.

Another significant challenge was governance. The governance of data and the use of data analytics tools was quite challenging for novice users. The issues of data quality and data integrity were also new to the management and they struggled in adequately addressing these issues.

Another major challenge was the lack of skills. Not only that the workforce was not tech-savvy, the managerial skills of the workforce were also limited to handle a project at such a mass scale.

I have mentioned the above challenges to give you a strategic direction that you should not make these mistakes or else your AI implementation will also have a huge risk of failure.

FIGURE 73: AI SUCCESS FRAMEWORK BY TALENTICAcxvii

Talentica has also presented a comprehensive AI framework for a successful AI implementation. According to this framework, the implementation should be carried out in different phases. In each phase, the tasks should be divided among data scientists, data engineers, and product engineers.

The first phase is business understanding where a comprehensive understanding will be developed of a particular maritime shipping company. Then the next phase is the acquisition of data. In this phase, data engineers will carry out the processes of data collection and data pipeline. Then data scientists will do the tasks of data clean up and data analysis.

The next phase is the development of AI model. In this phase, the data scientists will perform the tasks of feature engineering, model training, and model evaluation. The last phase is the deployment phase. In this phase, the tasks will be performed by product engineers. The product engineers will execute product integration, validation, and performance monitoring. As highlighted in the figure, this should be a continued process. So, for instance, if the validation is not successful, there will again be the activity of business understanding, and this cycle will go on.

Figure 74: System Engineering Loop for AIcxviii

Figure 75 shows that for a successful AI implementation, the system engineering cycle should not be considered as distinct from the conventional software development life cycle. The same processes are also involved in AI implementations. However, the conventional cycle is preceded by AI modeling cycle in AI-based implementations. In the modeling phase, the datasets are gathered and then explored. Then the relevant data points are integrated into the training system. As a result, a powerful AI model is developed, and this model is also evaluated by pilot testing.

At the bottom of Figure 74, the specific requirements of AI-based systems have also been indicated in each phase. In the modeling cycle, the required interfaces will be embedded AI and data quality metrics. These metrics will evaluate the size, quality, and relevance of the data. In the coding phase, the specific requirements are verification, validation, security, and integrity. In the deployment and release phase, the specific requirements are lower energy consumption, safety compliance, and adherence to ethical guidelines of AI use. If this paradigm of AI modeling is followed and the specific requirements are also considered, the chances of a successful AI implementation will substantially increase.

4.4. CALCULATION OF ROI FOR AI ADOPTION

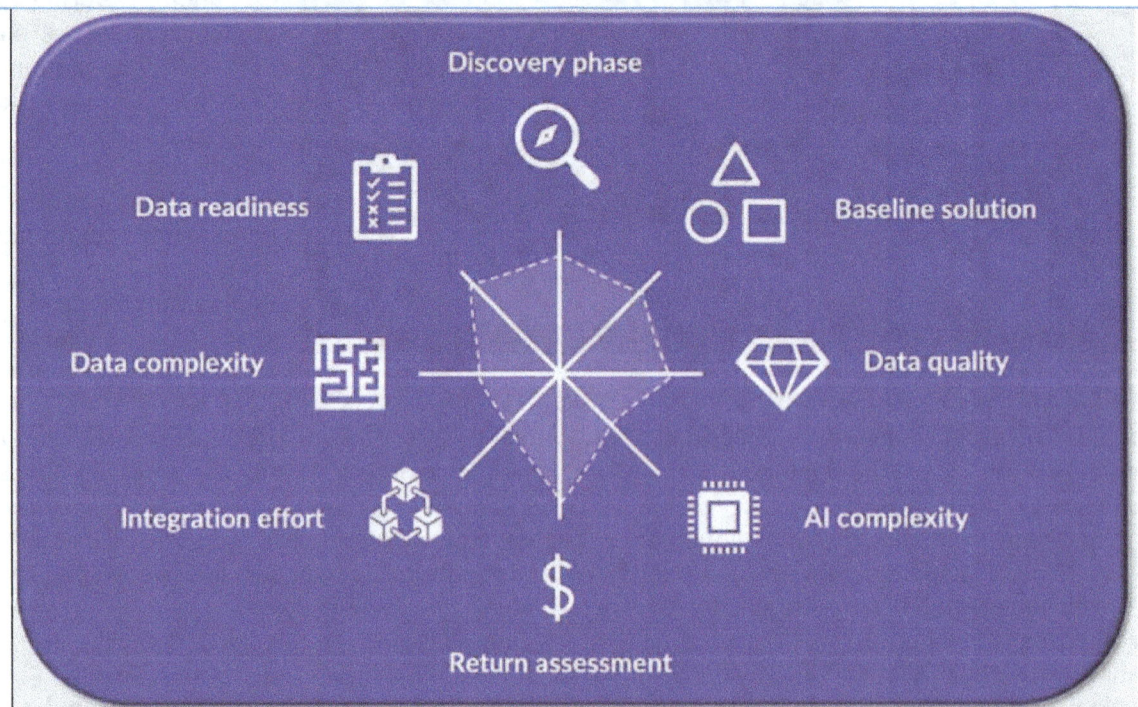

Calculating ROI for AI adoption becomes quite challenging because the ship owners, ship managers, maritime institutes and organizations do not possess the necessary skills to know the working of AI algorithm. As there is a process of learning and optimization also involved in the AI-based systems, calculating ROI in advance becomes quite complex. In this section, I will tell you several approaches and the best practices for calculating ROI.

4.5. FIGURE 75: RETURN ASSESSMENT FOR AI PROJECTScxix

Figure 75 mentions one approach for ROI calculation, and this approach is highlighted by medium.com. According to this approach, the ROI should be calculated right from the discovery phase to the return assessment. The first area of consideration is the data readiness. If the required datasets are not available, then it is a major negligence on the part of the ship owners and ship managers because datasets are the key inputs for the AI based model. The second factor to evaluate is the level of complexity in the

80

arranged data. There should be a consistency in the data and the combinations should not be so large enough that could affect the learning process of the AI model.

The risk managers should also evaluate the integration efforts because the interfaces are also required to be developed with the existing systems of the maritime shipping companies. The ROI can also be computed by considering a baseline solution. Then it should be evaluated by the risk manager how below or above a given AI system performed in the actual implementation phase. The data quality should also be a key performance indicator and it will reflect the efforts and investments made by the decision-makers. The last factor is the complexity of the AI algorithms. Although, a certain level of complexity is bound to arise in AI algorithms, the working of the algorithms should not be so complex that the results and recommendations become meaningless for the maritime shipping company.

4.6. FIGURE 76: FIVE IMPORTANT AI ROI METRICScxx

Dataiku has presented an AI ROI metrics in which five key indicators have been mentioned that should be calculated as part of AI and data science ROI as shown in Figure 76 above. The first indicator is the effect of ROI implementation on the revenues of the maritime shipping company. There should be a visible increase observed in customer conversions and the money spent on the shipment of cargoes. The second key indicator is the effect on costs. The cost savings should be seen from the perspective of both direct costs and indirect costs. The third key indicator is the competitive advantage. The services of the maritime shipping company should be clearly differentiated by the implementation of AI. The fourth key indicator is the speed to value. The project management team should be able to execute projects quickly after the availability of AI tools and technologies. The fifth key indicator is the team efficiency. The implementation should improve the productivity and efficiency of the team members, and as part of ROI calculation, it should also be estimated how the team efficiency resulted in the saving of costs for the maritime shipping organization.

The figure also highlights several other factors that should be part of ROI calculations besides the five key indicators. There should be no or much reduced risk of regulatory non-compliance in the new AI project. The cost of disruption should be within the capacities of the organization and there should be a sufficient availability of the funds. The monetary value of digital maturity should also be gained by the maritime shipping organization.

In this chapter, my focus was on describing the AI implementation plan steps, risk mitigation strategies, and ROI calculation. Now, let's move to the next chapter where I will elaborate the ethical concerns associated with the AI implementation.

5. ETHICAL CONSIDERATIONS IN IMPLEMENTING AI

Ethical considerations are the most overlooked area in AI based implementations and it can cost a lot to the maritime shipping organizations. The ship owners, ship managers, maritime institutes and organizations should be well aware of the ethical implications of their initiative or else they will have to face penalties and fines not only from the government agencies and the regulatory bodies, but the customers may also lodge complaints against those companies.

If we look at the broader level of the impacts of AI implementations, various ethical issues will be observed even at a first glance. Think of a chatbot that is answering all our questions in the form of ChatGPT and Google Bard. If the students pass their assignments and course work by using these chatbots, will it not be unfair for the students who are making genuine efforts? Think of a simple dataset processed by a conventional computer program versus large datasets utilized by AI programs for the learning and the development of data models. The high computational power of machines will increase the environmental footprint of the organizations. So, here is the ethical dilemma. On the one hand, we are using the AI systems for fuel efficiency and effective energy management. But the systems themselves are consuming a lot of energy and the maritime shipping companies might have to face the questions of the environmental protection organizations if there is a high consumption of energy by the processing of AI algorithms.

Another ethical issue emerges with the use of data. As I highlighted earlier, when the chatbots are answering your all questions, they are using the data resources on the web that may also be copyrighted. So, if you Google a resource, you will be shown the results as copyrighted. However, when you ask the same information from chatbots, they will give you the same information. So, it is a sensitive question that how the AI-based systems are acquiring the data and whether the data sources are legitimate or illegitimate. Moreover, the AI-based systems improve their performances by their interactions with the system. So, is it ethical to use the user generated data without the consent of the users and customers. These and other many ethical issues emerge as you look deep into the paradigm of AI implementation.

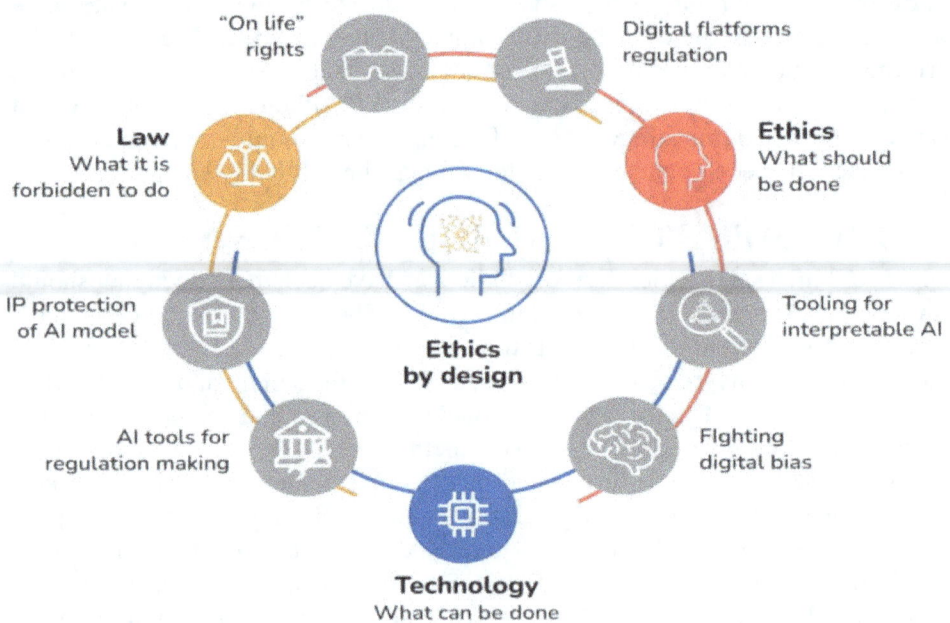

FIGURE 77: ETHICAL AI DESIGNcxxi

Figure 77 shows an approach of an ethical AI implementation. By following the above framework, the ethical approach could be embedded in the overall design of the AI implementation. In this context, the three pillars of AI implementation should be studied together that include the technology, the laws, and the ethics. From the regulatory perspective, it should be seen that what is allowed in the laws and what is prohibited during the AI implementation. The next area of consideration is to protect the AI model by creating a copyright of the model. It is essential because you will develop the data model after an extensive training and the learning process and a huge dataset of a very high quality will be fed to the AI systems. Therefore, it should not be easier for the other developers to copy your AI model in their implementations. The third aspect that can be beneficial to you is to acquire those AI tools and technologies that will facilitate you in the regulation making. It would be needed in cases where suppose you are pioneering a technology implementation in the maritime shipping organization.

The second aspect is related to ethics. For the ethical adherence, you should make an extensive research on digital platforms regulations and life protection rights. You might also be required to use explainable or interpretable AI so that the working of the AI algorithms could also be explained to the regulators and government agencies. You may also have to fight digital bias where the AI tools are favoring a particular course of action based on the design of the AI systems.

The third aspect is the technology. The technology should be used for an enhanced level of ethical adherence. When you are aware the extent to which the regulatory frameworks are to be complied with and the ethical guidelines are to be followed, it is the time to use technology for enforcing these guidelines.

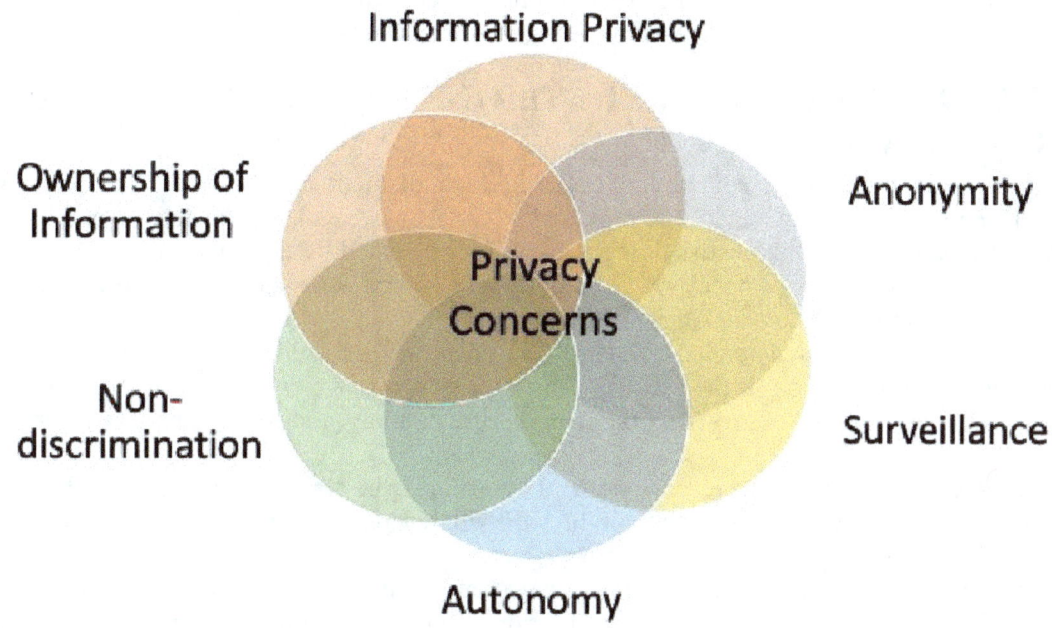

Figure 78: Data and Information Privacy Issues[cxxii]

Figure 78 highlights the data and information privacy issues in the ethical paradigm of AI implementation. Privacy concerns are raised because the AI algorithms are trained by acquiring data from multiple sources. The ethical question arises that who owns that data and whether that data can be legitimately used by the AI algorithms. Another issue that arises that I explained to you in the Amazon

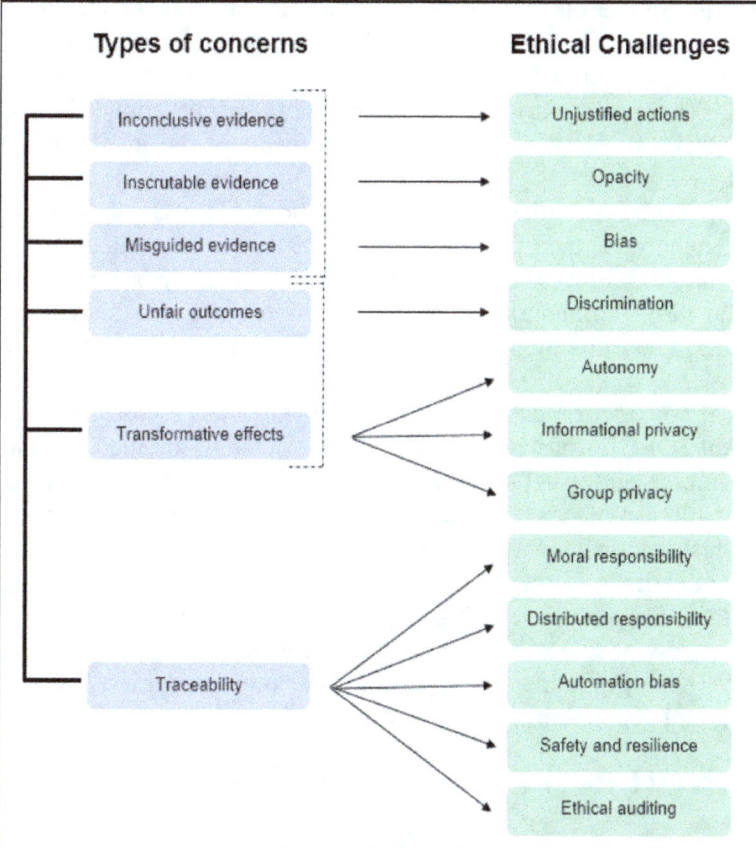

example is the discrimination inherently present in the dataset. For example, in the case of Amazon, when the dataset of old applications was provided to the recruitment robot, the robot incorrectly concluded that being male is a desirable attribute for being inducted at Amazon. It is because the majority of the old applications at Amazon had been submitted by the male candidates. The discrimination in data also raises the ethical question as to whether such data should be used for processing or not.

Another aspect is the anonymity of the data. When the data is processed by the AI algorithms, do they ensure anonymity in the sensitive details of the individuals or actual details are saved in the AI database. If actual details are saved, then this data may be extracted by an expert who knows how to run queries on the database. The surveillance mechanism of the AI system also raises ethical issues because the surveillance may spread to the sensitive areas. Suppose two countries are at war with each, then both countries will not want that their maritime regions are being monitored by the

AI systems. Due to this reason, the operations of autonomous drones are prohibited in various military areas because the drones can transmit sensitive information and details regarding those areas.

FIGURE 79: ETHICAL IMPLICATIONS BASED ON THE TYPE OF CONCERNcxxiii

Figure 79 above categorizes the ethical challenges in AI implementation based on different types of concerns. If there is inconclusive evidence provided by the AI systems, then the actions based on those recommendations will also be unjustified. If the evidence presented by the AI systems are inscrutable evidence, then the challenges of opacity will be observed. If the results are misguided due to the low quality of data, then the decision-making might be biased.

If unfair outcomes have been generated by the system, then the discrimination may be observed in the maritime shipping organization. If the results of the AI-based systems produce transformative effects, the ethical challenges will be observed in the context of autonomy, group privacy, and informational privacy.

If there are traceability issues in the results of the AI algorithms, the ethical challenges may be observed concerning automation bias, moral responsibility, distributed responsibility, ethical auditing, safety concerns, and resilience.

In this chapter, I have highlighted ethical considerations in AI implementation from multiple dimensions. First, I have discussed how the international regulatory frameworks govern the use of AI. Next, I have deliberated on the responsible use of AI. Then the environmental impacts of the AI implementations have been discussed extensively. Then I move to the sustainability considerations and the impact of AI on the job market and the job insecurity of the crew members.

5.1. INTERNATIONAL REGULATORY FRAMEWORK

Ethical implications of AI make a significant impact on how the ship owners, ship managers, maritime institutes and organizations should proceed with the implementation. The ethical concerns should also be addressed in the context of international regulatory framework because the maritime shipping operations are transnational operations. There are various guidelines already developed in the international regulatory frameworks for a responsible use of AI. The key aspects of the international regulatory framework are mentioned below.

The first and primary source of ethical premise is United Nations Guidelines. The UN has made extensive efforts for developing a comprehensive ethical framework so that all member states follow some standards while implementing AI. The guidelines of the UN were known to the general public in 2021. These guidelines highlight that the basic human rights and ethical considerations should be adhered to in all AI-based implementations. The implementations should not harm the interests of the stakeholders or the surrounding communities.

The second guiding framework for ethical considerations is OECD Principles. The Organisation for Economic Co-operation and Development (OECD) has also followed the footsteps of the UN and developed its own ethical guidelines for implementing AI systems. The guidelines emphasize that the business managers should make a responsible use of AI. It means that the AI should be utilized in the maritime shipping organizations by considering the factors of transparency, fairness, and accountability.

The third important resource on ethical guidelines is the UNESCO Recommendations. The United Nations Educational, Scientific and Cultural Organization (UNESCO) has also realized that the AI-based implementations come under the purview of scientific and educational initiatives, and therefore, ethical guidelines have been developed. These guidelines mention that the implementations should follow the principles of autonomy, equity, and accountability.

5.2. FIGURE 80: CORE PRINCIPLES OF ETHICAL AI BY UNESCOcxxiv

Figure 80 highlights the UNESCO recommendations for an ethical and responsible use of AI. The UNESCO regulatory framework is based on ten core principles and they have been termed as a human rights approach in the domain of AI. The first principle asks for not inflicting any harm by AI based implementations. The systems should not go beyond the admissible limits and the stated aims.

The second principle is that of security and safety. There should not be any unwanted harms caused by the AI systems and the security risks should be avoided to the extent most possible.

The third principle highlights that every individual and organization has a right to data protection and privacy. Therefore, the privacy should be requested at the moments of data acquisition and data processing by AI algorithms.

Other principles promote the stakeholder collaboration, responsible use, transparency, human oversight, sustainability, digital literacy, and fairness. It has been mentioned that while sourcing the data for training the algorithms, the national sovereignty and international laws should be respected. As per the principle 5, there should be traceability and auditability enabled in the AI based systems. The environmental wellbeing should also be ensured in the AI based implementations. As per the principle 6, the AI based systems should also have an interface of explainable AI so that the business managers know the rational of each recommendation and suggestion.

The UNESCO recommendations were presented in November 2021, and it was acknowledge in the general conference that AI ethics will be implemented by all 193 member states. As I highlighted above that the guidelines of the UNESCO framework focus on the incorporation of the values of dignity, human rights, diversity, inclusion, explainability, transparency, non-discrimination, fairness, responsibility, and accountability.

The basic premise of the UNESCO guidelines is the responsible use of AI. It is always a good strategy if the maritime shipping companies use AI for transforming maritime shipping for the future. However, as per UNESCO guidelines, the AI implementation should not harm the maritime ecosystem and protect the rights of the crew members and other workers. The implementations should also be sustainable because a huge dataset processing means using high power computer systems consuming a substantial energy resources. There should also be awareness and literacy regarding the use of AI tools, gadgets, and systems. The systems should not be used for unlawful activities such as harming the business activities of the competitors.

The UNESCO recommendations also highlight several areas of policy interventions. There should be an ethical impact assessment made by ship owners, ship managers, maritime institutes and organizations when they are engaged in AI implementations in their companies. An ethical governance model should also be formed so that the business managers are not only focused on expediting the implementation process, but also interested in knowing the ethical implications of those activities.

As per UNESCO recommendations, the organizations should also develop a data policy so that they could show the relevant documents and their business activities when there are queries and concerns regarding their acquisition of data. There should also be an international cooperation for implementing the ethical framework because the AI implementation is a new phenomenon for maritime shipping companies. It is only through the international collaboration, the ship owners, ship managers, maritime institutes and organizations can adhere to the ethical guidelines in implementing AI-based systems.

The UNESCO recommendations have been regarded as a landmark achievement in the domain of ethical AI framework because a global framework has been introduced by this document. The implementation of this tool can ensure a responsible use of AI tools, systems, and technologies. It can become a valuable resource for ship owners, ship managers, maritime institutes and organizations for transforming maritime shipping for the future.

Another source of ethical guidelines is the European Union (EU) Regulations. As I mentioned earlier, the formulation of a variety of regulatory frameworks may also create some issues for ship owners, ship managers, maritime institutes and organizations. For example, if there are EU regulations, and at the same time, there are also UK-based regulations or the IMO regulations, then the maritime shipping organizations may have to follow conflicting guidelines. In some of the cases, there is a more enabling environment and the procedural details have been left to the shipping companies such as UK-based regulations. However, in the other cases, there are strict regulations and the micro-level management. The numerous regulatory frameworks at the international level necessitate that a consistent ethical framework is developed at the global level for AI-based implementations. EU has developed an AI act that provides the data privacy laws as well as the protocols for AI-based implementations.

European Guidelines are also supported with data protection regulations known as GDPR. EU guidelines also address the issues such as the transfer of data outside the member states of the European Union. According to these regulations, the ownership and control of the data lie with the residents and citizens. Therefore, the international businesses will have to follow the EU guidelines even if the data centers are located outside the EU region.[cxxv] This amendment has been made by replacing the earlier 1995 directive of data protection. The new regulations are enforced from May 2018.

One of the unique features of GDPR related regulations are that the ethical guidelines are applicable to all business entities that process data within the EU region. It means that even if the organizations are headquartered outside EU but if the data processing occurs within EU, the organizations will have to comply with the GDPR regulations. It is a significant clause in the EU guidelines because maritime shipping organizations perform operations that touch many ports and terminals. Therefore, the adherence to EU guidelines will be highly crucial to the maritime shipping companies while implementing AI.

The EU guidelines make it very evident that as part of the ethical guidelines, the data owners (individuals and organizations) have various rights over their personal data. The personal data should also be accessible to them and they should also be able to modify their personal details. If an individual feels that a highly confidential and personal information have been exposed to the datasets, the individual should have the opportunity to erase or delete that data. The individuals and organizations should have the opportunity that they could restrict the processing of their data by AI algorithms. They should also be given the freedom to object to the use of their data by the AI algorithms.

GDPR guidelines also impose various restrictions on the organizations that implement AI systems and use the datasets for training the AI algorithms. In those cases, the ship owners, ship managers, maritime institutes and organizations should obtain an informed consent from the individuals and organizations whose data is being used for processing. The organizations should also implement technical and security measures for protecting the personal data and sensitive information. Even after implementing all the security protocols, if a data breach occurs in the systems implemented by maritime shipping organizations, the data breach should be reported to the supervisors within a duration of 72 hours. The maritime shipping organizations should also create a post of Data Protection Officer if the volume of dataset is large enough and the datasets contain sensitive, personal information of the individuals and the organizations.

The GDPR regulations have not just been introduced as a nice to have ethical guidelines. The maritime shipping organizations are required to comply with these regulations. In the case of non-compliance, the ship owners and ship managers will have to face a fine for non-compliance. The penalties may amount to up to 4 percent of the global annual turnover of the maritime shipping organization or €20 million. The greater value between these two amounts will be charged as the penalty. GDPR regulations are complex in the sense that maritime shipping organizations have recently embraced AI in their business domains and the ship owners and ship managers are not well aware of the legal implications of the AI implementation. However, it is crucial and mandatory that the ship owners, ship managers, maritime institutes and organizations process personal and organizational data by considering the EU regulatory framework so that they could meet their obligations under the GDPR framework.

Another notable contribution is the Global AI Governance Initiatives. Under this initiative, the ethical guidelines have been developed through international collaborations. The collaborations have resulted in AI governance initiatives, such as the Global Partnership on Artificial Intelligence (GPAI). These ethical guidelines are useful and beneficial for experts and industry leaders to address AI ethical challenges.

Another initiative that has gained a huge prominence is the Sector-Specific Guidelines. The sector specific guidelines for implementing AI have also been developed by IMO for the maritime shipping industry. These guidelines are focused on the responsible use of AI. These guidelines are highly beneficial and relevant for ship owners, ship managers, maritime institutes and organizations for enhancing security, safety, and environmental sustainability. The uniqueness of these guidelines is that they consider the specific scenarios of the maritime shipping operations.

Another notable global initiative is the Responsible AI Development. Under this initiative, an international framework has been developed. The framework highlights the significance of responsible AI development that considers all the ethical dimensions of embracing AI. This includes preventing AI from being used for those motives that are harmful to the humans or exacerbate biases in the industrial and commercial sectors.

A key dimension of ethical AI is the Transparency and Accountability. When ethical AI is used in the organizations, there is a requirement of transparency in decision-making processes. The systems should be conceived and developed in a way that provide explanations for their decisions so that it is possible to hold the individuals accountable in the case of unexpected happenings.

Another crucial aspect in the ethical paradigm is Bias Mitigation. International regulations have been developed with a great level of focus on mitigating biases in AI algorithms. The issue of bias should also be evaluated in the context of training data. The AI systems should be developed in a way that discriminatory outcomes could be avoided.

A key ethical dimension is the notion of Human Rights and Privacy. The basic human rights should be respected and the privacy of the individuals and communities should be considered when implementing AI. Organizations should ensure the compliance of AI applications with international regulatory framework on human rights and data protection.

The ethical premise of AI implementation also applies to Ethical AI Education and Training. When the crew members are trained for using AI-based systems, the international framework encourages education and training programs and it is a widely accepted practice. However, the training of the users should also raise awareness of best practices and ethical considerations.

Another key aspect in the ethical paradigm is the Ongoing Monitoring and Adaptation of the AI based systems. The international regulatory framework has a wider level of acknowledgment that AI algorithms evolve and there is a continuous process of improvement. Therefore, the ethical considerations must also be adapted and the guidelines regarding the new challenges should be added. Maritime shipping companies are encouraged to monitor their AI implementations and update the ethical guidelines to meet the current ethical challenges.

The above examples highlight the key global initiatives that have been made regarding the ethical challenges in AI implementations. When the ship owners, ship managers, maritime institutes and organizations adhere to this international regulatory framework, the maritime shipping industry will be able to capitalize on the opportunities offered by AI. At the same time, their initiatives will be aligned with human values and ethical principles. This responsible use of AI will promote safety, trust, and accountability in the overall maritime shipping ecosystem.

5.3. RESPONSIBLE USE OF AI

When there are refinements in different AI models and there is a wide scale implementation of AI in the maritime shipping industry, there will be temptations among the ship owners, ship managers, maritime institutes and organizations to spread the use of AI in all business processes and systems. However, as I cautioned above, the use of AI should be with the ethical boundaries and comply with the international and the local regulatory framework. This notion is now popularly known as the 'responsible use of AI'.

AI is enabling transformation in industries, including maritime shipping industry. The ethical considerations will also be more significant over time and the industry professionals will demand the responsible use of AI. The maritime shipping organizations should prioritize ethical guidelines and always ensure that AI technologies are developed and implemented in a way that is consistent with societal well-being and moral values. Below, I have highlighted key dimensions that should be considered in respect to the responsible AI use:

The first area of consideration is Fairness and Bias Mitigation. The AI developers should ensure that AI algorithms are developed and trained to be non-discriminatory, fair, and unbiased. The inherent biases in training data should be addressed to prevent unjust treatment of groups. For example, if the dataset favors a certain ethic group, gender, or age group, then the skewed nature of the data should be critically evaluated by the statistical experts and the data should be cleaned up for consistency and normalization.

The second factor of consideration is Transparency and Explainability. AI algorithms should provide clear explanations as to why they are recommending a particular decision or action. The managers cannot allow a particular decision without knowing themselves the significance of a particular direction. All the stakeholders should be able to understand the working of AI algorithms and why AI arrived at specific suggestions and recommendations, particularly in safety-critical scenarios. For this purpose, the ship owners, ship managers, maritime institutes and organizations might have to use the tools of explainable AI or interpretable AI. These tools will add to the cost of implementing AI. However, the ship owners and ship managers cannot overlook the significance of these tools in providing the rationale of their choices.

The third dimension is Accountability and Liability. The ship owners, ship managers, maritime institutes and organizations should clearly define and allocate accountability for AI-based implementations and their decisions. If there are errors, accidents, or collisions based on AI decision making, then the ship owners and ship managers should establish mechanisms for identifying responsible individuals and addressing the liabilities related to those incidents and collisions.

The fourth area of concern is Data Privacy and Consent. The ship owners, ship managers, maritime institutes and organizations should adhere to data privacy regulations both national and international. They should obtain informed consent when collecting, storing, and processing personal data for training the AI algorithms and developing the AI model. They should ensure that AI applications protect the rights of all individuals and sensitive information is not disclosed to unintended recipients.

The fifth dimension is Ethical Considerations in Design. The AI developers should not adopt an approach where the AI system is first developed and then the ethical considerations are evaluated. The ethical considerations should be embedded in the design phase of AI systems. The development of the objectives, goals, and capabilities should be in line with the ethical guidelines. The AI developers should also consider the potential impacts of the AI systems and interfaces on society.

Another crucial dimension is Human Oversight and Control. As I have explained at various points in this book, the AI based systems should not be seen as a threat to the jobs or the replacement of human interventions. The AI based systems are used only to complement the efforts of the humans and ensure more accuracy and precision in the decision-making. The ship owners, ship managers, maritime institutes and organizations should still maintain human oversight of AI systems. It is particularly essential in critical domains such as maritime safety. They should always ensure that the crew members can intervene and override the AI decisions when the ground realities necessitate such interventions.

One crucial aspect in the responsible use of AI is Avoiding Harmful Use. The ship owners, ship managers, maritime institutes and organizations should prevent the use of AI for harmful or malicious motives. They should hire AI experts that could develop AI applications with safeguards such that the chances of misuse or abuse are minimized.

Another aspect in AI based implementation is Human Dignity and Non-Discrimination. While adopting the mechanistic approach, the data should be processed by AI algorithms by respecting the dignity and fundamental rights of the data providers. AI developers should avoid discriminatory practices and the use of AI in the maritime shipping should be aimed at promoting inclusivity and equality.

Another dimension regarding the responsible use of AI is the Continuous Monitoring and Auditing. The ship owners, ship managers, maritime institutes and organizations should regularly monitor and audit AI systems. The ethical violations, biases, and unintended consequences should be identified and rectified. They should ensure the implementation of those mechanisms that could report ethical concerns.

Another key dimension for responsible use is Ethical Education and Training. The ship owners, ship managers, maritime institutes and organizations should develop a culture of ethics and responsible AI within the maritime shipping company. The training and education regarding AI tools should be provided to employees that emphasizes the significance of ethical AI development and implementation.

Another core dimension regarding the responsible use is Ethical Impact Assessment. The ship owners, ship managers, maritime institutes and organizations should conduct ethical impact assessments. Only when the satisfactory results are obtained in the ethical assessments, the AI systems should be deployed in in critical roles. The ship owners and ship managers should evaluate potential ethical, social, and environmental outcomes, and refine the deployment strategy as needed.

Another core dimension is Public Engagement and Transparency. The AI implementations hit multiple data centers and may also affect other individuals and communities. Therefore, a good level of collaboration is a key to a successful AI implementation. The ship owners, ship managers, maritime institutes and organizations should engage with the public, stakeholders, and the communities to discuss AI implementation. They should listen to potential concerns and ethical considerations. Throughout the implementation, the focus should be on promoting transparency and accountability in the decision-making process of AI implementation.

The last dimension regarding the responsible use of AI is the Collaborative Ethical Frameworks. If you develop the ethical frameworks specific to your organizations, your framework may conflict with other maritime shipping companies and you will struggle in the implementation in the maritime regions. Therefore, the ship owners, ship managers, maritime institutes and organizations should collaborate with industry peers, international organizations, and governmental bodies to formulate and adhere to ethical frameworks that guide the responsible use of AI in the maritime shipping industry.

In the above section, I have mentioned 13 core dimensions regarding the responsible use of AI. The ship owners, ship managers, maritime institutes and organizations will have realized by this description that prioritizing responsible AI use is not only a moral obligation but also for the organizational reputation and the long-term success of AI applications in maritime shipping. When ship owners, ship managers, maritime institutes and organizations embrace ethical guidelines, they can realize the true benefits of AI while reducing risks and promoting a culture of responsible innovation.

5.4. ENVIRONMENTAL IMPACT

The analysis of the environmental impact also comes under the purview of the ethical considerations. It is because if the maritime shipping companies are involved in embracing AI for gaining a competitive advantage, they cannot do so at the cost of harming the environment and consuming more energy resources.

The growth of AI-based systems has also resulted in the development of the systems that are highly connected and integrated. Therefore, they may have a higher negative impact on the environment when the impacts are considered in totality with all connectivity and the use of energy resources. Ethical considerations concerning environmental sustainability are crucial to analyze for the ship owners, ship managers, maritime institutes and organizations. This analysis will ensure that AI contributes positively to the sustainable business as well as a sustainable environment. Below, I have mentioned the key areas of ethical environmental considerations for implementing AI-based systems.

The first area of consideration is the Energy Efficiency. AI systems should prioritize energy optimization as a key agenda. Energy-intensive AI processes can result in an enhanced environmental harm and enhanced carbon emissions. The maritime shipping companies should aim to develop and deploy AI systems that could be operated with minimal energy.

The second key area is the Carbon Footprint. The ship owners, ship managers, maritime institutes and organizations should measure the carbon emissions and reduce the carbon footprint of AI systems. They should assess the emissions generated during the development, training, and operation of AI models. Moreover, the energy consumption of AI hardware should also be calculated.

The third critical area is a Sustainable Hardware. The hardware solutions should be environmentally sustainable for the development of AI infrastructure. The infrastructure development should involve selecting energy-efficient data centers. The utilization of renewable energy sources should be preferred. The ship owners, ship managers, maritime institutes and organizations should also implement cooling solutions that minimize energy usage.

Another area of consideration is the Green Data Centers. The ship owners, ship managers, maritime institutes and organizations should select data center providers that implement the principles of green data center, such as minimizing energy consumption and the utilization of cooling systems. This holistic approach will make a substantial contribution to a reduced environmental impact.

Another critical area of consideration is the Environmental Impact Assessments. The ship owners, ship managers, maritime institutes and organizations should conduct environmental impact assessments. These assessments should be carried out before deploying AI systems in maritime shipping. They should evaluate the potential impact on the environment, including air quality, marine ecosystems, and the utilization of non-renewable resources.

Another critical environmental factor is the Sustainable AI Applications. AI developers should design AI applications that foster the notion of sustainability within the maritime shipping industry. For instance, they should promote the use of AI to optimize ship routes and minimize fuel consumption, and minimize emissions. In this way, they will have an active role in developing a more sustainable shipping operation.

Another key area of consideration is the Recycling and E-Waste Management. When the implementations are based on the concepts of Ethical AI, the ship owners, ship managers, maritime institutes and organizations should ensure a responsible disposal of AI hardware and electronic waste. They should implement e-waste management and recycling practices to reduce the environmental impact of outdated AI gadgets.

Another critical environmental factor is the Sustainable Data Practices. The ship owners, ship managers, maritime institutes and organizations should introduce sustainable data practices by reducing data storage needs, adhering to the policies of data retention, and utilizing efficient techniques of data compression.

In the environmental domain, there is an important concept of Carbon Offsetting. The ship owners, ship managers, maritime institutes and organizations should implement carbon offsetting strategies to compensate for the emissions that are inevitable in the AI-based operations. Based on this strategy, the maritime shipping organizations are establishing the targets of net zero carbon emissions where the emissions will be offset by other sustainability initiatives. Therefore, the ship owners and ship managers should invest in initiatives that reduce an equivalent amount of carbon from the business operations.

A key environmental decision is the Data Center Location. When the AI based systems are implemented, the ship owners, ship managers, maritime institutes and organizations should select data center locations that align with their sustainability agendas. They should locate data centers in regions with access to renewable and clean energy sources to minimize the carbon footprint of AI systems operations.

As I mentioned earlier, the Regulatory Compliance is also another key area. The ship owners, ship managers, maritime institutes and organizations should adhere to environmental regulations and standards relevant to AI based implementations. They should ensure compliance with environmental laws and consider it as an ethical responsibility for sustainable AI implementation.

Another key area of consideration is the Environmental Impact Reporting. The sustainability reports are periodically published by multinational organizations and these reports should also mention the impact of AI based operations. The ship owners, ship managers, maritime institutes and organizations should transparently report the environmental impact of AI based implementations to regulators, stakeholders, and the general public. When they provide data on energy consumption, emissions, and sustainability efforts, the maritime shipping companies can be held accountable for the environmental impact of their business operations.

The last area of consideration is the Collaboration for Sustainable Solutions. Environmental impacts can be reduced more effectively through collaborative efforts. The ship owners, ship managers, maritime institutes and organizations should collaborate with governmental bodies, environmental organizations, and industry experts to formulate best practices for sustainable AI implementation in maritime shipping companies.

In the above section, I have mention various key areas of consideration for reducing the environmental impact of AI based operations. Ethical considerations concerning the environmental impact are becoming increasingly significant, and maritime shipping companies have a responsibility to ensure that AI technologies have a positive contribution to sustainability targets. When the ship owners, ship managers, maritime institutes and organizations implement sustainable practices, transparently report environmental impact, and reduce energy consumption, the maritime shipping industry can achieve an eco-friendly use of AI.

5.5. SUSTAINABILITY CONSIDERATIONS

The responsible use of AI is a broader paradigm and the ethical adherence of a business entity is viewed in conjunction to the sustainability considerations. Sustainable AI adoption in the maritime shipping companies can address social and environmental concerns and the real benefits of AI-based implementations can be achieved. Below, I have mentioned various areas where the sustainability considerations should be evaluated in the context of AI implementation.

The first core area is the Environmental Sustainability. The ship owners, ship managers, maritime institutes and organizations should evaluate the the environmental impact of AI implementations. They should use energy-efficient hardware and data centers so that the carbon footprint of AI implementations could be reduced. They should select green data centers, renewable energy sources, and efficient cooling systems as I highlighted comprehensively in the previous sections.

The second area of consideration is the Carbon Emissions Reduction. The ship owners, ship managers, maritime institutes and organizations should use AI to optimize operations in maritime shipping. The implementations should minimize fuel consumption and emissions. AI should be utilized in developing more fuel-efficient routes, managing energy consumption on ships, and optimizing maintenance schedules. All these efforts will contribute to a sustainable maritime shipping industry.

The third area of consideration is an Eco-Friendly Data Management. The maritime shipping companies should implement sustainable data practices. The ship owners, ship managers, maritime institutes and organizations should use the techniques of data compression, efficient data storage, and data center location choices that are aligned with the sustainability agenda. They should reduce the environmental impact of data management in AI based systems.

Another sustainability consideration is the use of Green Data Centers. The ship owners, ship managers, maritime institutes and organizations should build partnerships with green data centers that give preference to sustainability. These data centers utilize energy-efficient systems, renewable energy sources, and cooling systems to reduce the carbon emissions related to AI based implementations.

One more sustainability factor is the Responsible Hardware Lifecycle. The ship owners, ship managers, maritime institutes and organizations should address the entire lifecycle of AI hardware. They should consider the lifecycle from production to disposal. They should introduce recycling and e-waste management practices to reduce the environmental impact of obsolete AI gadgets.

Another area of consideration in sustainability is the Environmental Impact Assessment. As I also mentioned in the previous section, the ship owners, ship managers, maritime institutes and organizations should conduct environmental impact assessments. These assessments should be made before deploying AI systems in maritime shipping companies. They should evaluate the potential effects on the environment, such as air quality, marine ecosystems, and the use of non-renewable resources.

Another critical area of consideration is the Marine Conservation. The ship owners, ship managers, maritime institutes and organizations should use AI for sustainable management and marine conservation. AI based systems should not harm the marine life and the systems should monitor and manage fish populations, prevent overfishing, and protect marine ecosystems. All these efforts will contribute to sustainable maritime practices by using AI based systems.

Another crucial sustainability factor is the Ethical Data Sourcing. The ship owners, ship managers, maritime institutes and organizations should ensure that data used in the development of AI models, such as marine environmental data, is sourced by adhering to the ethical principles. The data acquisition should respect international regulatory framework.

A key sustainability factor is the Human and Social Sustainability. The ship owners, ship managers, maritime institutes and organizations should introduce AI systems with a focus on social sustainability. They should evaluate the effect of AI on crew members and communities in the maritime shipping industry. They should also ensure that AI-based automation does not negatively affect local economies or employment opportunities.

Another sustainability consideration is the Community Engagement. The ship owners, ship managers, maritime institutes and organizations should engage with the communities where the AI based systems are to be deployed. They should also collaborate with local stakeholders and consider their economic, social, and environmental issues so that there is a sustainable integration of AI with the existing systems.

Another key dimension in the context of sustainability is the Sustainability Reporting. The ship owners, ship managers, maritime institutes and organizations should transparently report on the impacts of AI implementations and sustainability efforts. They should provide data on energy efficiency, emissions reduction, and other initiatives concerning environmental and social sustainability.

The last area of consideration in sustainability is the Collaboration for Sustainable Solutions. The ship owners, ship managers, maritime institutes and organizations should build collaborations with governmental bodies, industry peers, environmental bodies, and other stakeholders to formulate best practices. As a result, there will be a sustainable AI implementation in maritime shipping companies.

In the above section, I have presented a comprehensive list of sustainability areas. As is evident from my description, the sustainability considerations are as crucial and important as the ethical considerations. The business entities cannot overlook the environmental impact of their business operations and the AI based implementations will form an integral part of the sustainability reporting of maritime shipping companies. When the ship owners, ship managers, maritime institutes and organizations incorporate sustainability considerations into AI adoption, the maritime shipping companies will be able to transform maritime shipping for the future. The will realize the transformative potential of AI and also contribute to a more sustainable and responsible future of maritime shipping companies. The sustainability considerations highlight that there should be a balanced economic progress in maritime shipping companies. The social and environmental responsibilities are also crucial for developing a harmonious and sustainable maritime future.

5.6. IMPACT OF AI ON MARITIME INDUSTRY'S JOB MARKET

As I highlighted in my discussion on the cultural challenges associated with AI implementation that the AI based implementations also create a sense of job insecurity among the population. The crew members tend to believe that AI implementations will eat their jobs and more and more robots will replace the humans. The senior management in those organizations should make the workers realize that AI implementations are for improving the quality of output and enhancing the efficiency. The knowledge and experience of the hired workers will still be respected and the implementations will not result in downsizing or retrenchment.

The ship owners, ship managers, maritime institutes and organizations should realize that AI implementations can improve operational efficiency and safety. However, it may also result in workforce disruptions. Therefore, they have a responsibility to address this ethical concern. It is crucial to ensure a responsible and balanced transition to the AI based systems. In the below section, I have highlighted several areas of consideration in this regard.

The first critical area is Job Displacement and Reskilling. The ship owners, ship managers, maritime institutes and organizations will come to know that as AI and automation technologies become more prevalent in their organizations, there will be chances of job displacement, particularly in repetitive, routine, and manual tasks. They have an ethical responsibility to invest in upskilling and reskilling programs for affected workers. It will ensure that these workers are still retained in the workforce and their experience is utilized for more strategic level tasks and responsibilities.

The second people management factor is the Social Responsibility. The ship owners, ship managers, maritime institutes and organizations should demonstrate social responsibility by ensuring that the AI based implementations will not result in mass layoffs. They should develop alternate strategies with the help of the HR department such as voluntary attrition, job redeployment, job enrichment, or offering early retirement options.

The third critical area in people management is the notion of Equity and Inclusivity. The ship owners, ship managers, maritime institutes and organizations should realize that ethical AI implementation means they will have to respect equity and inclusivity. They should ensure that the benefits of AI are shared across the workforce. They should promote diversity and inclusion in AI development, integration, and deployment phases.

Another area of consideration is the Human Oversight. As I also explained earlier, the ship owners, ship managers, maritime institutes and organizations will always require human oversight in AI-based processes, particularly in maritime safety-critical tasks. Human interventions should remain in control and the humans should be allowed to intervene in AI-based systems if required.

Another people management factor is the notion of Hybrid Workforce. The ship owners, ship managers, maritime institutes and organizations should foster the concept of a hybrid workforce. In such a workforce, the AI augments human capabilities and is not deployed with the intent of replacing them. AI should be used for handling repetitive tasks, freeing up human resources to focus on decision-making approaches and more strategic level tasks.

Another ethical factor in the context of people management is the Early Warning Systems. The ship owners, ship managers, maritime institutes and organizations should develop early warning systems so that the workforce could identify potential job displacement and workforce disruptions. The business managers should also take proactive measures to minimize these challenges, including training and workforce reallocation.

Another people management perspective is the Labor Market Analysis. The ship owners, ship managers, maritime institutes and organizations should periodically assess the labor market to predict the growing demands for skills and roles in the maritime shipping companies. The ship owners and ship managers should utilize this information to guide training and educational programs for maritime workforce.

Another people management area of consideration is the Labor Unions and Negotiations. The ship owners, ship managers, maritime institutes and organizations should build collaborations with labor unions and worker representatives in negotiating fair and equitable adoption of AI based systems. The participation of workers in discussions regarding the impact of AI systems on their jobs is an ethical imperative and will reduce the concerns and worries of the workforce.

Another people management perspective is the Public Awareness and Education. The ship owners, ship managers, maritime institutes and organizations should raise public awareness about the impact of AI on the maritime job market. They should prepare the workforce and the wider community about the possible changes in the job dynamics and the steps needed to improve their skill set.

Another area of consideration is the Government and Regulatory Oversight. The regulatory bodies and governments play a crucial role in framing the ethical implementation of AI in the maritime shipping companies. They should develop ethical guidelines that address workforce concerns and queries. These guidelines should promote a responsible use of AI.

Another area of consideration is Retaining Institutional Knowledge. The ship owners, ship managers, maritime institutes and organizations should recognize the value of institutional knowledge and memory in the maritime shipping industry. They should consider measures to retain the expertise of experienced employees and use their corporate memory in improving the efficiency and effectiveness of maritime shipping operations.

The last area of consideration is the Sustainable Job Creation. The ship owners, ship managers, maritime institutes and organizations should realize that AI implementation can also create new job opportunities in fields related to maintenance, AI development, and implementation oversight. They should promote the creation of sustainable jobs to balance the losses of the current routine-based jobs.

The ethical implications should be considered by the maritime shipping professionals regarding the job market. The ship owners, ship managers, maritime institutes and organizations should focus on an inclusive and creative approach in this regard. They should focus on the well-being and equitable treatment of the maritime shipping professionals. The industry can realize the full potential of AI by addressing the mentioned ethical challenges.

On the one hand, AI-based interventions are creating job insecurity among the workforce because the business processes are going to be automated. However, on the other hand, AI-based technologies are also opening a new door of business ventures in the technological domain. The AI-based startups are in high demand these days and there is a reduced barrier to entry in this market. It is because the AI developers are available in limited numbers. Therefore, if the startups offer new and innovative solutions, they are highly welcomed by the business community.

The technology professionals also make it a point when the ethical implications are posed to them concerning the impact on job market. The supporters of AI argue that AI has also opened several new job opportunities. If the maritime shipping professionals also learn these new technologies, they can improve their employability potential manifold.

Below, I have mentioned the examples of five machine learning vendors that have achieved remarkable success by introducing machine learning technologies in the maritime sector.

Figure 81: Soshianest, First Success Storycxxvi

Figure 81 shows the first example of a successful startup in the maritime sector introducing machine learning tools and technologies. Soshianest is a Canadian-based startup and offers shipping rate prediction solutions based on machine learning technologies. The maritime historical data is used for predicting the rates of shipment. The mining platform makes use of an AI-based metaheuristic algorithm. The model has a proven prediction accuracy and the competitive environment is also considered in the prediction of shipment rates.

Figure 82: Blockshipping: Second Success Storycxxvii

Figure 82 highlights the second example of a successful startup in the maritime sector introducing machine learning tools and technologies. Blockshipping is a Denmark-based startup and offers AI tools for the maritime shipping based on machine learning technologies. The system is focused on the optimization of containers' processing at the terminals. The AI system is used for calculating the dwelling time of the container. Then, the optimized arrangement of the containers is recommended based on the dates of shipping. The system has a proven track record of improving throughput capacity.

FIGURE 83: VAKE: THIRD SUCCESS STORYcxxviii

Figure 83 mentions the third case of a successful startup in the maritime sector introducing machine learning tools and technologies. Vake is a Norway-based startup company. The company has developed an AI-based geospatial intelligence platform that can detect the position of a ship in the sea through the satellite. Various satellite images are processed and a high level of classification is achieved by using machine learning technologies. The platform is not only beneficial for the maritime shipping companies but also provides useful insights to the intelligence agencies and coast guards. The ship traceability enables the ship owners, ship managers, maritime institutes and organizations to predict the arrival time with a high level of accuracy.

FIGURE 84: ALPHA ORI: FOURTH SUCCESS STORYcxxix

Figure 84 highlights the fourth example of a successful startup in the maritime sector introducing machine learning tools and technologies. Alpha Ori is a Singaporean-based startup company offering machine learning-based solutions. The solution offered by the company is known as SmartShip. This AI-based solution monitors the entire operations of the vessel and tracks the emissions of the ship. Then the recommendations are made regarding the optimized fuel consumption and reduced carbon emissions. The sustainable shipping operations can be ensured by the ship owners, ship managers, maritime institutes and organizations by using this AI-based SmartShip platform.

FIGURE 85: BLUEPULSE: FIFTH SUCCESS STORYcxxx

Figure 85 shows the fifth example of a successful startup in the maritime sector introducing machine learning tools and technologies. Bluepulse is a French-based startup specializing in machine learning-based solutions. The company offers a data analytics platform for the predictive maintenance of the vessel. The AI-based system generates automated reports regarding the energy consumption, downtime, system failures, and the maintenance requirements of different machine parts and equipment. It is a brilliant solution for the ship owners, ship managers, maritime institutes and organizations for ensuring the safety of the vessel and reducing the costs associated with the system failure.

In the above section, I have given you five key examples of the startups that have shown an impressive growth in their businesses by offering maritime AI solutions based on machine learning technologies. Earlier, I highlighted to you ethical aspects and considerations that may emerge during AI systems implementation in the maritime shipping companies. The purpose of highlighting these five examples in this section is to make you realize that the proponents of AI still have a very optimistic view regarding the benefits of using AI-based systems. An advanced learning of the usage and development of these systems have also enabled the individuals to build their own startups and offer their customized solutions based on their skills and knowledge. Therefore, the ship owners, ship managers, maritime institutes and organizations should also be hopeful regarding the potential of AI in transforming maritime shipping for the future. I have mentioned challenges and ethical implications for giving you awareness on these areas and ensuring your journey a successful one. I hope that the description of the five successful cases will have further motivated you regarding the incorporation of AI-based tools in your maritime shipping companies. In the next chapter, I will discuss about the future trends in AI for maritime shipping and the road ahead. I will introduce to you the emerging AI-based technologies suitable for maritime shipping. I will also highlight the tools and systems available concerning the predictive maintenance. The disruptive technologies have also been mentioned next. You will find the next chapter highly insightful for viewing the future landscape of the transformation of maritime shipping companies by using AI tools and systems.

6. FUTURE TRENDS AND ROAD AHEAD

6.1. EMERGING TECHNOLOGIES BASED ON AI

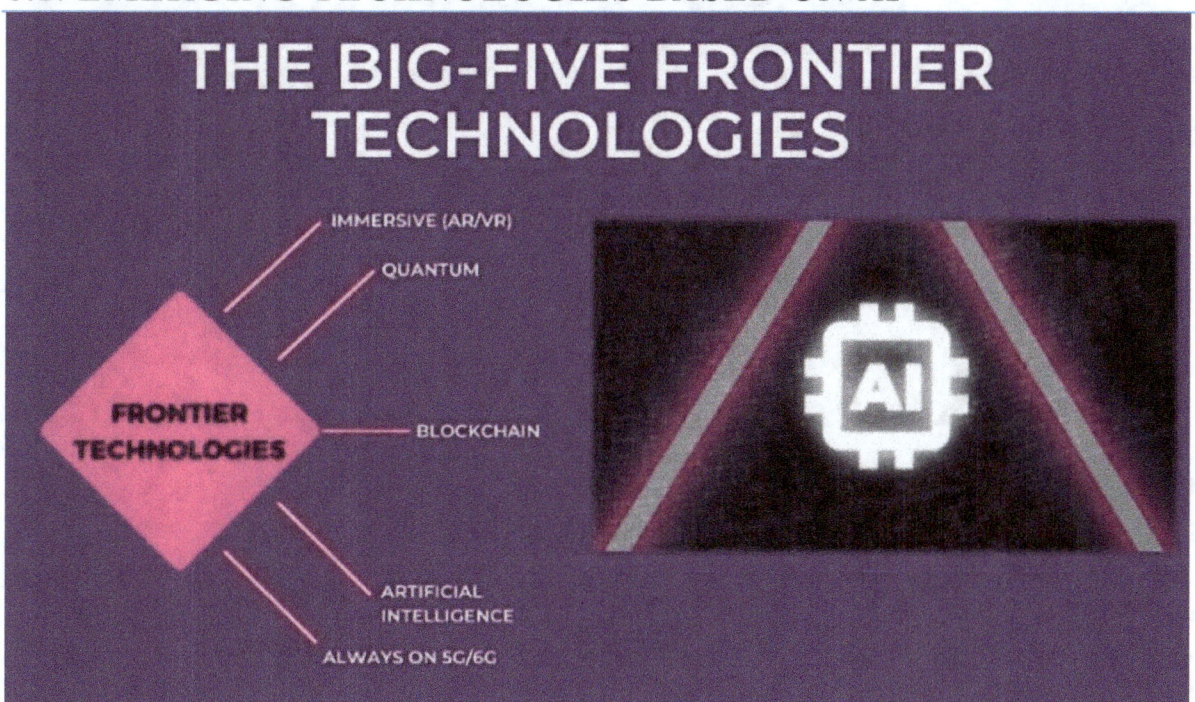

Emerging technology is a term that is used for a technological innovation that is about to be introduced or an improvement in the already introduced technology. Some of these technologies are known as frontier technologies because they are enabled by both scientific breakthrough and real-world implementations. TransformBase has mentioned five such technologies as shown in Figure 86 below:

FIGURE 86: FRONTIER TECHNOLOGIEScxxxi

As shown in the figure above, artificial intelligence is also regarded as one of the frontier technologies. The blockchain based systems and quantum computing systems also implement the concepts of artificial intelligence. In this section, I will tell you those emerging technologies that are expected to be introduced in the near future and that have the potential to transform maritime shipping for the future.

Figure 87: Emerging Technologies in Maritime AIcxxxii

Figure **87** shows the top trends in maritime AI that will transform maritime shipping for the future. As is evident from the figure, the use of artificial intelligence is at the top of the list. These trends have been presented by StartUs by analyzing the future directions of 1,163 maritime organizations. The figure also shows the leading vendors and business entities in the maritime shipping industry that are embracing AI. These trends make it imperative for the ship owners, ship managers, maritime institutes and organizations to unleash the potential of AI and gain a competitive advantage in the maritime shipping industry.

FIGURE 88: IMPACT OF THE TECHNOLOGICAL INTERVENTIONS ON MARITIME ECOSYSTEMcxxxiii

Figure 88 highlights the results of the data analysis carried out by StartUs regarding the impact of technological interventions in the maritime shipping sector. The figure endorses that artificial intelligence will have the most significant impact (26%) on the business processes of the maritime shipping industry. The figure also mentions that the ship owners, ship managers, maritime institutes and organizations should also consider other areas of intervention such as clean energy, energy efficiency, robotics, and IoT. All these features can be incorporated and integrated in the AI-based systems.

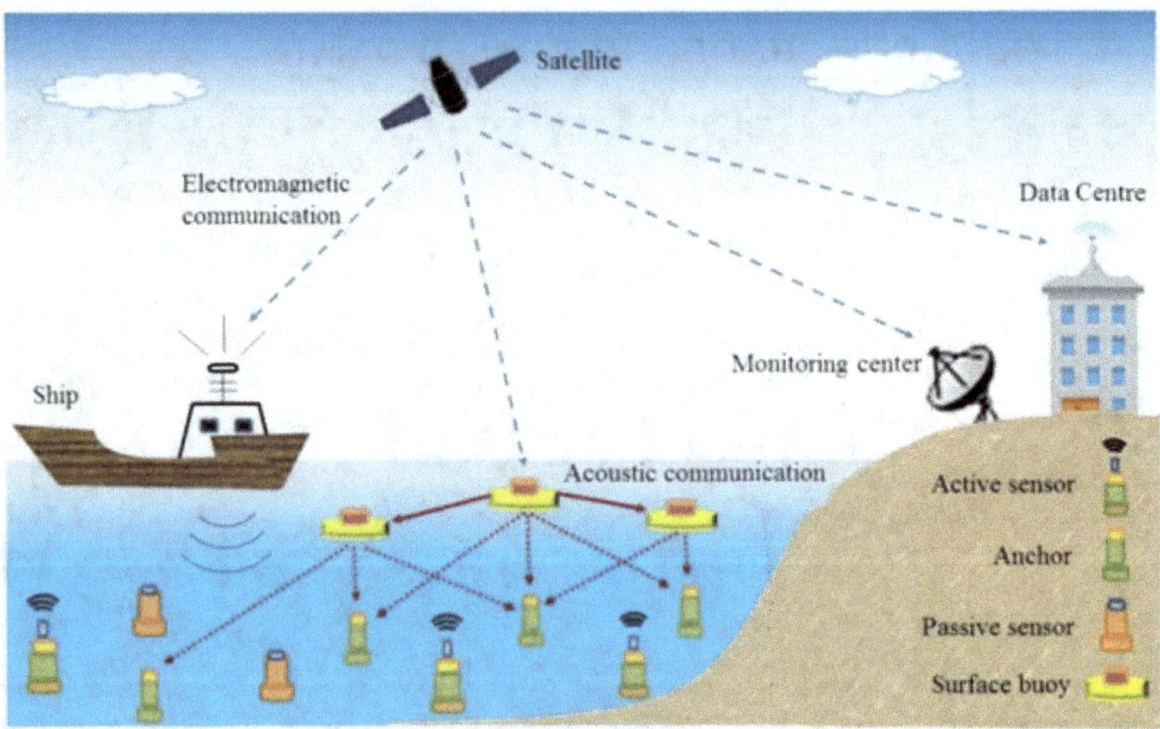

FIGURE 89: INTERNET OF UNDERWATER THINGS (IOUT)cxxxiv

Figure 89 shows a new technology paradigm in maritime AI that has been introduced by MDPI in a journal article. By following the analogy of Internet of Things (IoT), this paradigm has been termed as Internet of Underwater Things (IoUT). This framework is highly significant for the ship owners, ship managers, maritime institutes and organizations because internet connectivity in the maritime region will pose new challenges and issues when AI-based systems will be implemented in the maritime shipping. As shown in the figure, the ship owners will have to enable connectivity of the ship through the satellite. This connectivity can be made possible by electromagnetic communication. The monitoring center will also receive data via satellite. The offshore data center should also be connected with the satellite so that there are no wired connections affecting the receiving of high-speed internet bandwidth.

The acoustic communication should be enabled by considering various parameters and factors. There should be surface buoy directly connected to the satellite. They should be able to differentiate between active sensors and passive sensors. The data should be collected from the active sensors by the surface buoy. The anchors should act as facilitators in the data transmission between surface buoy and active sensors. This proposed model can make a significant improvement in the current connectivity models of AI-based systems that are operating in the maritime shipping industry. The processing of huge datasets by AI algorithms is possible only if a high bandwidth internet connectivity is available to them. This connectivity can be accomplished effectively by using the model of IoUT.

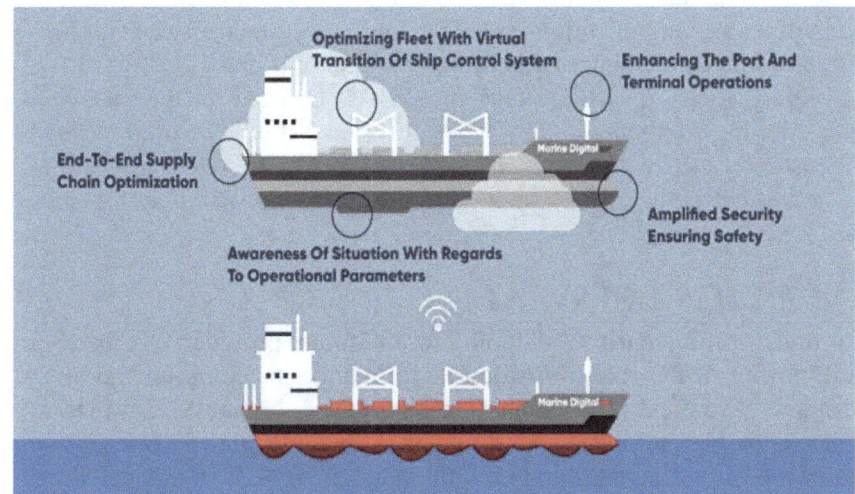

FIGURE 90: USING DIGITAL TWINS IN MARITIME AIcxxxv

Figure 90 presents another application of AI that can benefit the entire maritime ecosystem. The AI-based implementations are also benefitting from the concept of digital twins. In the case of a digital twin, physical system and machines are represented in digital forms. Then simulations can be run by ship owners and ship managers in the virtual environment. When the operations are performed through a combination of physical and virtual worlds, there is more effective analysis of data and the effective monitoring of the systems prevents undesired outcomes. When the concept of a digital twin is introduced in the maritime shipping sector, there will be a broader understanding of the trade patterns. The digital twin will bloom the operational as well as strategic decision-making process. This concept can also optimize the port operations and terminal operations because it will address the issues of punctuality, berth allocation, and cargo spacing requirements. The systems will have more situational awareness in relation to the operational parameters. The systems can also move forward to the end-to-end optimization of the supply chain. The security and safety parameters of the maritime operations can also be strengthened by introducing digital twins in the AI-based systems.

FIGURE 91: AI INTERVENTIONS FROM THE SHIPMENT PERSPECTIVEcxxxvi

Figure 91 highlights how the AI interventions can be beneficial from the shipment perspective. Maritime shipping is a combination of the maritime operations and the shipment operations. There are also various logistics arrangements involved outside the sea area. Therefore, the ship owners, ship managers, maritime institutes and organizations can realize the full benefits of AI if they focus on its implementation not only from the perspective of maritime AI but also form the lens of shipment and logistics AI.

From the perspective of shipment AI, Figure 91 shows various emerging technologies based on AI. As far as the planning process is involved, the AI tools and technologies can be used for demand forecasting and supply planning. For the development of an automated warehouse for maritime shipping, the ship owners and ship managers can use warehouse robots, the predictive maintenance systems, and damage detection systems. The damage detection systems are highly significant for the maritime shipping because the goods and the cargo should be delivered to the customers in a good condition and if there are risks of damages, there should be systems in place that could report the risks to the ship owners, ship managers, maritime institutes and organizations.

Another area of AI applications in maritime shipping is autonomous things. Under this paradigm, the maritime shipping companies can use the delivery drones and self-driving vehicles to ship the product to the customer's location once the product arrives at a particular port and terminal. Another significant intervention in the logistics domain is data analytics. The AI analytics tools can facilitate the ship owners, ship managers, maritime institutes and organizations to introduce dynamic pricing mechanism. The analytics systems can also be used for route optimization.

Another key AI intervention is in the area of back office operations. The shipment process through the maritime route also involves an extensive documentation process so that the prohibited goods could not be smuggled from one place to another. Therefore, the AI tools should also be used for enabling automation in the document processing. The manual office tasks should also be automated. The ship owners, ship managers, maritime institutes and organizations should also use customer service chatbots so that the regular business operations could be executed without human interventions.

In the paradigm of logistics and shipping, the AI-based systems should also be used for the sales and marketing activities of the maritime shipping companies. The AI-based systems should be used for lead scoring and highlighting potential customer profiles and customer segments. The systems should also be used for automating the tasks of email marketing. The systems can also be used effectively for sales analytics and marketing analytics.

6.2. PREDICTIVE ANALYTICS USING AI

Predictive analytics can be used for transforming maritime shipping for the future provided that the ship owners and ship managers are fully aware of the role of artificial intelligence in the optimization of maritime shipping's business operations. Frontiers Science has explained the role that AI can play in the form of a pyramid that resembles to the hierarchy needs pyramid as shown in Figure 92 below. The pyramid shows the transformation of the business beginning with the manual work. The manual work is improved by the use of Data. The data utilized in this context could be raw data, measures, and video footages. Then the systems further evolve, and the data is converted into information. The information includes the processed outputs and analyzed outputs by the AI algorithms. The next development stage reaches when the information is converted into knowledge. At this stage, there are the formation of concepts and theories. The maritime shipping companies build collaborations and share their ideas and success stories. The last and the peak stage is that of wisdom. When a maritime shipping company reaches this stage, there is a higher level of understanding of both the business processes and the systems. The crew members have actionable targets and goals. The decision-making process is aimed at achieving successful business outcomes. Therefore, the whole benefits of automation and artificial intelligence are achieved in the form of an increased synthesis of data and converting data into wisdom.

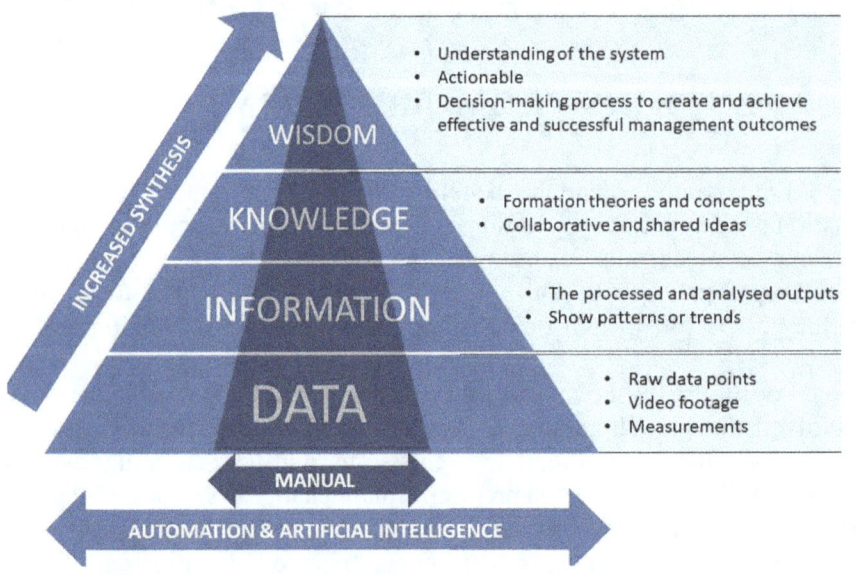

Figure 92: Path from Manual Work to AI-Based Automationcxxxvii

The AI-based predictive analytics should also be viewed in the same concept of converting data into wisdom. The machines and equipment are attached with sensors in an AI-based system. These sensors provide real-time information regarding the health status of the machine and equipment. If there are deviations of the machines from the set standards, the alerts are sent to the data centers. As I also explained in the previous sections, the AI-based systems can be used in the predictive analytics only to send alerts and provide dashboard indicators or they can also be automated to correct the errors identified in machines and systems. This auto-correction of machines and equipment can be highly expensive in the initial stages of AI-based implementations for the ship owners, ship managers, maritime institutes and organizations. Therefore, in the first phase, the ship owners and ship managers should use the predictive analytics only to view the alerts and health statuses. The auto-correction can then be implemented at the later stages of implementation.

The data analytics has become a highly advanced field after the implementation of AI-based systems. As this section is devoted to highlighting the future trends, I want to highlight that the world is now moving even beyond predictive analytics and the new paradigm has been termed as prescriptive analytics.

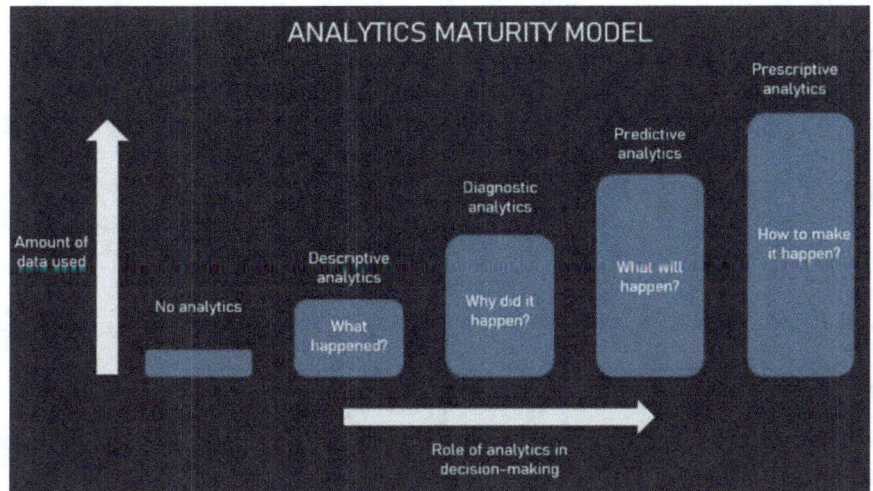

Figure 93: The Maturity Evolution of Analyticscxxxviii HeavyAI has explained the maturity levels and evolutions of the analytics nicely through a diagram as shown in Figure 93 above. As shown in the figure, the use of an analytics model depends on the amount of data used. If you have a very limited data to process, why would you go for acquiring advanced predictive analytics systems?

Therefore, based on the above figure, if the data to be processed is very small, then the ship owners and ship managers can even be comfortable with using the conventional computer systems and applications. If there is a slight growth in the database, then you can use descriptive analytics systems that will tell you what happened at different touchpoints of maritime shipping and you will get a holistic picture through these systems. If you are interested in gaining more sense of your data and also want to know the causes of certain system behaviors, then you should go for diagnostic analytics systems. Predictive analytics is a stage next to diagnostic analytics. It is a stage where a maritime shipping company has implemented AI tools and technologies. As a result, the AI algorithms and machine learning technologies are processing huge datasets from various systems and interfaces. The use of such heavy processing should result in meaningful insights for the business entity. Therefore, in the next stage, you will be required to install predictive analytics systems. These systems will highlight to the ship owners, ship managers, maritime institutes and organizations what is going to happen during the vessel voyage and shipment operations. These key insights will be useful for both the business managers and the customers. Once the predictive analytics systems are installed and used successfully in a maritime shipping company, your next goal should be to use prescriptive analytics systems. These systems will simply make you realize your dreams. They will show you an exact path and roadmap based on data processing as to how you can convert your plans and strategies into actions.

FIGURE 94: BUSINESS IMPACT OF PREDICTIVE ANALYTICScxxxix

Figure 94 highlights the business impact of using predictive analytics tools. The illustration assists in comprehending how the predictive and prescriptive analytics based implementations can transform maritime shipping for the future. The illustration shows the business impact from transactional to strategic on x-axis. The engagement maturity has been drawn from low to high in the illustration. The basic data services should be used by the ship owners, ship managers, maritime institutes and organizations when only a transactional level impact is to be reported. The value-added services should be used when the organizations are slightly moving towards a strategic orientation. The use of predictive analytics is regarded as the third phase, and it has been labelled as the phase C in the illustration. In this phase, the ship owners, ship managers, maritime institutes and organizations should use predictive analytics as well as prescriptive analytics systems. In this context, predictive modeling and natural language processing will provide a low level of customer engagement. If the customer insights modules are used, then a middle level engagement can be expected. The higher level engagement can be achieved when propensity models and forecasting models are used. The next and the highest level of achievement for the maritime shipping companies will be to use turn-key data services. In this context, the middle level engagement can be achieved by using machine learning and deep learning technologies. The highest level of engagement can be accomplished by implementing cognitive-based automation and AI-based automation.

While implementing predictive analytics models, it is also crucial for the ship owners, ship managers, maritime institutes and organizations to know how AI-based predictive analytics systems work. Each system and vendor offer a unique set of features, however, the basic working of predictive analytics systems has been explained by process.st.

The Predictive Analytics Process

Pull	**Prepare**	**Pick**	**Predict**	**Plan**
Extract the data from where it lives.	Clean, refine, and prepare it.	Identify what to predict.	Create the prediction.	Develop a plan of action.

FIGURE 95: HOW PREDICTIVE ANALYTICS SYSTEMS WORKcxl

Figure 95 highlights the dynamics of AI-based systems. First, the system extracts the data from the multiple sources from where the data acquisition has been designed for the AI model. Then, the system refines the data and prepares it in the format that is suitable for the data analysis. Then, the system focuses on various queries and parameters for which predictions are to be made based on the available and processed data. In the next stage, the prediction is created by using sophisticated machine learning algorithms. Then the predictions are shown to the users and a plan of action is also attached with the recommendations. This working is highly useful for the ship owners, ship managers, maritime institutes and organizations because the maritime shipping operations are highly costly and incorrect decision-making can result in massive losses. There is a need for introducing data-driven decision-making and the tools of predictive analytics can enable this feature in the maritime shipping operations.

FIGURE 96: USE CASES OF PREDICTIVE ANALYTICScxli

Figure 96 mentions various use cases of predictive analytics. These uses of predictive analytics indicate the road ahead and the potential of the predictive analytics in transforming maritime shipping for the future. Predictive analytics systems are based on the tools of machine learning. The machine learning technologies identify key patterns in the data. Then they score the significance of various data attributes based on their relevance to the prediction queries. The data visualization is also achieved in the system for providing greater visibility of the data to the end users. The predictive behavior of the system is always available to the business managers in real-time. The decision-makers can then opt for data-driven decision-making. The predictive analytics models also facilitate smooth communication among all stakeholders. The statistical modeling can also be developed in the system. The data insights are also related to the evolving consumer behavior that can assist maritime shipping companies in updating the portfolio of their services.

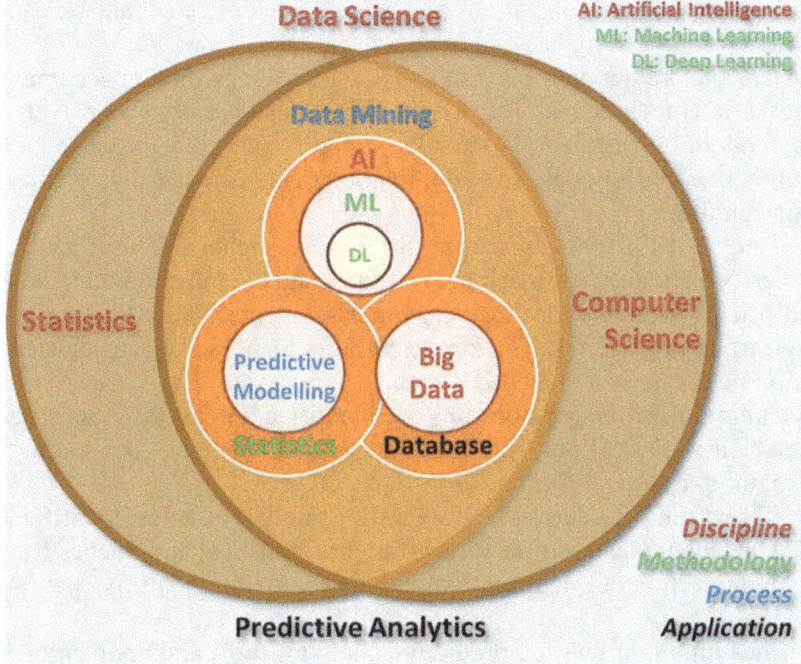

Figure 97: Future Models of Predictive Analyticscxlii

Figure 97 highlights that the future models of predictive analytics will be a unique blend of data science and predictive analytics. All the models will be developed at the intersection of statistics and computer science. Within this intersection, the key goal of predictive analytics systems will be the mining of the big data. From the AI perspective, machine learning and deep learning algorithms will be used for this purpose. From the statistical perspective, the predictive models will be used for predicting the future state of the business operations. From the database perspective, the system should be capable for storing and processing big data. Through different coloring schemes, the figure also highlights that artificial intelligence is a discipline, whereas predictive analytics is an application for making predictions based on the processed data. Therefore, while presenting the AI-based solutions to the ship owners, ship managers, maritime institutes and organizations, the AI developers and the IT team should provide clarity to the business people regarding the differentiation between discipline, methodology, process, and application.

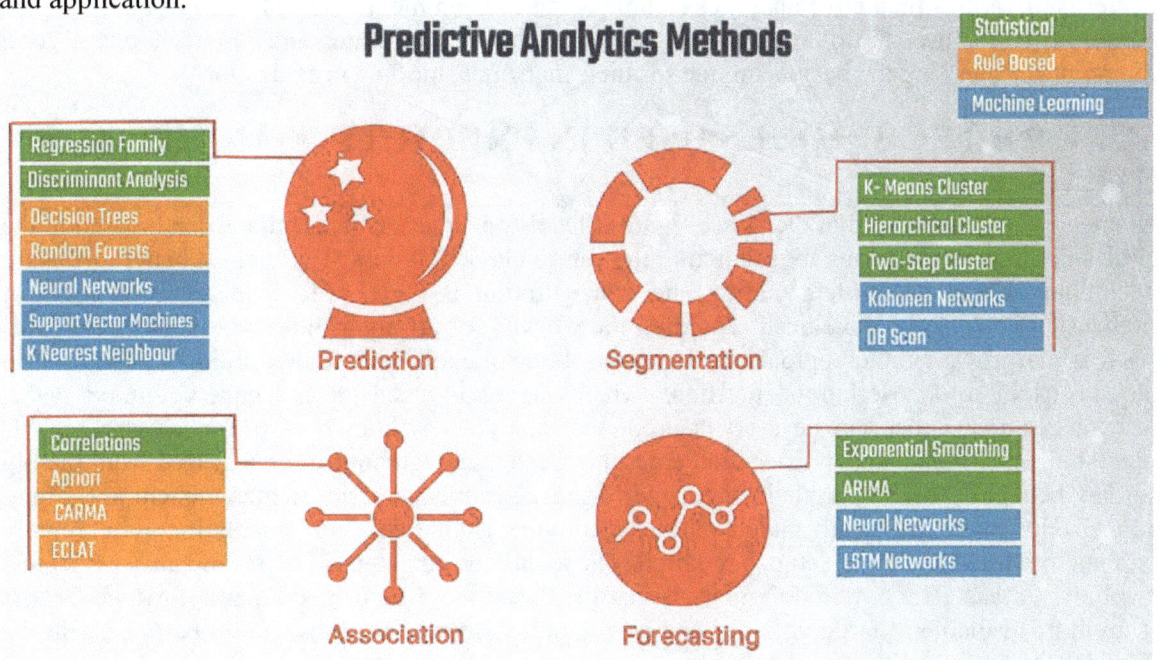

FIGURE 98: FUTURE APPROACHES TO PREDICTIVE ANALYTICScxliii

Figure 98 highlights the four key approaches that will lead the future landscape of predictive analytics. The ship owners, ship managers, maritime institutes and organizations should evaluate which of these approaches can provide maximum benefits to their organizations and how a competitive advantage can be achieved by using AI-based, predictive analytics tools.

The first of these approaches is known as the prediction tools. In this approach, the prediction accuracy is improved and optimized by using various techniques. The AI developers use regression analysis, discriminant analysis, decision trees, random forests, neural networks, support vector machines, and k-nearest neighbor. As shown in the figure, these techniques are a combination of statistical analysis, rule-based analysis, and machine learning techniques.

The second approach is known as segmentation. In this approach, different segments and clusters are identified from the data. The ship owners, ship managers, maritime institutes and organizations can then introduce customized incentives and pricing schemes for those customer segments. These techniques are based on statistical and machine learning algorithms. The key techniques include k means cluster, hierarchical cluster, two-step cluster, kohonen networks, and DB scan.

The third approach is known as association in which relationships are identified among different data attributes. These techniques are based on statistical and rule-based algorithms. The key techniques include correlation analysis, apriori method, camra, and eclat.

The last approach is known as the forecasting models. The forecasting models use the identified associations for making predictions. The techniques used for forecasting are statistical approaches and machine learning techniques. The main techniques include exponential smoothing, arima, neural networks, and lstm networks.

The description of these four approaches highlight that the predictive analytics tools and applications have evolved significantly and are using advanced statistical, rule-based, and machine learning approaches. It is a significant future trend in the AI-based implementations. The ship owners, ship managers, maritime institutes and organizations should select those approaches and techniques that could provide the biggest advantage in their organizational context. As is evident from my description above, the maritime shipping professionals may find it hard to select the right approach and technique of predictive analytics. They will have to acquire the expertise of IT experts, AI developers, and other technology professionals to select the best course of action. In Figure 99 below, the purpose, merits, and demerits of various predictive analytics techniques have been mentioned. My description on these techniques will give the ship owners, ship managers, maritime institutes and organizations a good starting point for selecting the best technique for their maritime shipping organization.

FIGURE 99: TECHNIQUES USED IN PREDICTIVE ANALYTICS SYSTEMScxliv

The first technique highlighted in the above figure is Decision Trees. This technique can be used in the maritime shipping organizations for predicting the future classes of data. The biggest benefit of using this technique is that its implementation and configuration is easier. Moreover, a good level of comprehension can also be developed regarding the working of AI algorithms. The drawback of this approach is that it is a simple approach and may not be applicable in complex problem solving. The technique is useful for answering the questions 'which one' and 'yes/no' from a huge volume of data.

The second technique that can be used in predictive analytics is known as neural networks. This technique builds clusters and classifiers and emerging customer segments are recognized from the big data. A key benefit of this technique is that its performance is even superior to many machine learning algorithms. However, the disadvantage of this technique is that the implementation is extremely difficult and parallel processing setup is required. The technique can be used to get any answers from a large volume of data. The only requirement is that the data should be of good quality. Moreover, the data should be available in large volumes and correlations and relationships should be present in the supplied data.

The third technique is known as regression analysis in which both linear regression and logistic regression are used. This technique uses the existing relationships among the variables to predict the future outcomes for the maritime shipping organization. A key benefit of this approach is that the statistical results are delivered to the users that can easily be understood and interpreted. A limitation of this approach is that the studied data should clear the assumptions of linear regression and logistic regression. This technique assists in highlighting the key target variables to the business managers.

The fourth technique is known as time series. Through this technique, continuous or scale variables are forecasted over time. The benefit of this approach is that the results are intuitively clear and easy to understand. However, the drawback of this approach is that it requires a dataset, which is time-dependent. This technique evaluates the current state of the data and predicts the future state of the data. The fifth important technique is known as clustering by using k-means. This technique can be used for organizing the data into groups based on the similarity. The benefit of this approach is that implementing this technique is comparatively easy. However, a big challenge in this approach is that the k number of clusters is to be predicted first. This technique is useful in identifying the available patterns in the data. The last crucial technique is known as factor analysis. This technique is used to evaluate the variables and based on their contributions in explaining the data, the variables are converted into several key factors. The benefit of this technique is that it is a dimension reduction technique and provides a concise, representative picture of the data. However, the interpretation of the factors may become complex, and key information might be lost while building the factors. This technique provides good explanations regarding the emerged themes from the data.

From the explanation of these six techniques of predictive analytics, you will have got a basic idea how different techniques can be utilized in the real-world scenarios. You should devote sufficient time for selecting the techniques and algorithms to be used in AI systems. As I explained, all of these techniques have their benefits and limitations. A wrong choice may reach you to a point where the backtracking is not possible for you considering the costs and resources involved in the project management.

6.3. DISRUPTIVE TECHNOLOGIES

In the domain of information and communication technology, those technologies are considered as disruptive technologies in which innovation is at a highly advanced level such that there is a paradigm shift in the way businesses, consumers, and industries are operating at present. From the ancient ages to the present, wheel, the bulb, and the mobile phones can be categorized as disruptive technologies because they entirely changed the way businesses are being conducted globally. The ASEAN Post has mentioned the list of 12 disruptive technologies as shown in Figure 100 below:

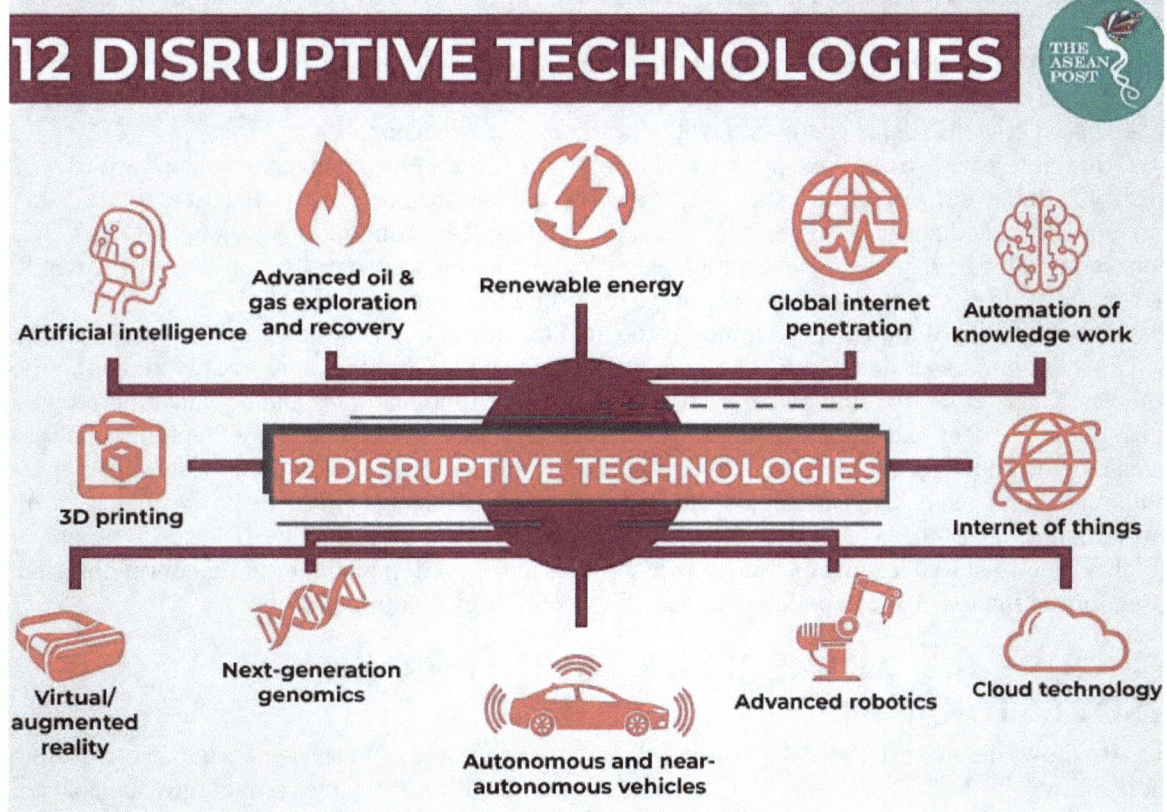

6.4. FIGURE 100: EXAMPLES OF DISRUPTIVE TECHNOLOGIEScxlv

Figure 100 shows several examples of disruptive technologies. As can be seen, artificial intelligence is at the top of the list. Its key relationship is with the automation in the knowledge society. Based on my explanations in this book, you will have noticed that AI also has relationship and dependence on other disruptive technologies such as global internet penetration, Internet of Things, cloud technology, advanced robotics, and augmented reality.

6.5. FIGURE 101: MARITIME SECURITY AREAScxlvi

Figure 101 mentions various ways and means through which disruptive, AI-based technologies can be used in the maritime shipping for improving the security in the maritime area. Achieving security in the maritime region is a highly desired feature for the ship owners, ship managers, maritime institutes and organizations because the AI-based systems are threatened not only by the physical threats but also the cyber security threats. There is also a huge data exposure to the AI-based systems and it is the responsibility of the ship owners and ship managers to maintain the confidentiality and privacy of the data.

The figure highlights various external factors that influence the security of a ship. Therefore, the security experts should not only focus on the security of the vessel but also the external factors influencing the ship operations.

First of all, there are crew members on-board the ship and they are using WiFi and wired connections on their laptops and smartphones. These gadgets can be accessed by the adversaries for tracking the location of the vessel. Therefore, the security should be strengthened in the crew network. The ships are also connected with the satellite systems, VHF, and GMDSS. The data transmission protocols should be highly secure in this communication mode. The vessels may also have an emergency position indicator that generates alerts and alarms at time of a threat. As I explained earlier, if the AI systems do not receive a good quality of data, the alert systems may also lose its significance.

The vessels are also connected with radar systems, position finding systems, and data recorder systems. The security of this network arrangement should also be strengthened by AI tools and systems.

The vessel operators also have a computerized passenger information system and the profiling of the passengers can also be optimized by the AI-based systems. The rescue services and evacuation services in the security management should also be improved by AI-based systems. The vessels use a variety of navigation aids such as echo sounder, ECDIS, AIS, and radar systems. These systems can also be moved to the AI-based systems for an improved level of accuracy and precision.

The weather forecasts are the key pieces of information for a ship manager because various key decisions for the vessel voyage are taken after reading the weather forecasts. AI-based systems can provide up-to-date and accurate weather forecasts by fetching data from various weather stations. The systems should also be capable of reporting iceberg locations, hurricane tracks, and the drift patterns. The AI-based systems can also generate warnings regarding the severe storms.

Another crucial aspect of the vessel operation is the loading and stability of the vessel. There have been incidents where the vessel was overloaded with passengers and several valuable lives were lost in the maritime area. Therefore, the ship owners, ship managers, maritime institutes and organizations should also ensure the efficient loading and stability of the vessel. In the figure above, Bay Planning Software has been recommended for this purpose. Other AI-based systems should also be used by the ship owners and ship managers such as ballast systems and hull stress monitoring systems. The power management systems and machinery management systems can also be optimized by using the AI-based systems.

StartUs has identified five leading disruptive technologies in the AI domain that can transform maritime shipping for the future. I have explained all these five technology solutions below:

6.6. FIGURE 102: AI-BASED SHIP MONITORING SOLUTIONScxlvii

Figure 102 shows the first AI-based disruptive technology mentioned by StartUs. It is a ship monitoring solution offered by Metis Cyberspace. The system uses machine learning technologies to analyze carbon emissions and fuel consumption of the vessel voyage. The system provides valuable insights for improving the operational efficiency of the vessel systems.

6.7. FIGURE 103: AI-BASED COLLISION AVOIDANCEcxlviii

Figure 103 shows the second AI-based disruptive technology mentioned by StartUs. It is an AI-based collision avoidance system during the vessel voyage, and is offered by Orca AI. The platform keeps checks on all the fleets so that the fleet managers could assess the risk of collisions. The hazards can also be detected by the crew in congested waterways and complex port operations.

6.8. FIGURE 104: AI-BASED AUTONOMOUS SHIPPINGcxlix

Figure 104 mentions the third AI-based disruptive technology mentioned by StartUs. It is an AI-based autonomous shipping system and it is also supported with vessel situational awareness. The system is developed by Captain AI, which is a Dutch startup. The system makes an efficient utilization of AI-based simulations, due to which autonomous ships can also operate at high seas. The lower fuel consumption and carbon emissions are enabled by the utilization of neural networks. The AI-based systems also ensure a high level of prediction accuracy.

6.9. FIGURE 105: AI-BASED CONTAINER INSPECTIONcl

Figure 105 shows the fourth AI-based disruptive technology mentioned by StartUs. This system is developed by Canscan, which is a Canadian startup company. As I explained in the previous sections, transforming maritime shipping for the future means that the ship owners, ship managers, maritime institutes and organizations will need to transform maritime-based business operations as well as shipment-based business operations. Therefore, container inspection automation can prove to be a revolutionary, disruptive technology for the maritime shipping industry. The system easily integrates with the existing camera devices and highlights damages and other issues. The containers are also evaluated against the safety standards and the non-compliance is immediately reported. In the area of predictive maintenance, these machine-learning based technologies can ensure an effective asset maintenance in the maritime shipping industry.

Figure 106: AI-Based Risk Analyticscli

Figure 106 mentions the fourth AI-based disruptive technology mentioned by StartUs. The risk management is a highly crucial aspect in the ocean area and AI-based ocean risk analytics can be a highly beneficial aspect for the ship owners, ship managers, maritime institutes and organizations. This system has been developed by a US-based startup known as Scoot Science. The system develops a trend analysis based on abnormal oceanic events because the primary objective of the system is to conduct a risk analysis. The trends are presented in the form of a dashboard known as SeaState. The insights from this AI-based system enable the ship owners and ship managers to know about the potential risks well ahead and then risk mitigation strategies can be developed accordingly.

The description of these five disruptive technologies mentioned by StartUs highlight that AI vendors are targeting different aspects of maritime businesses to introduce AI-based innovations. The state-of-the-art systems have been developed for the collision avoidance, autonomous vessels, ship monitoring, container inspection, and ocean risk analytics. Now as a ship owner, ship manager, or a maritime institute, you should evaluate those areas that are in dire need of optimization through AI. You should also evaluate the strategies of your competitors as to which areas they are targeting for AI-based innovations. Then, you should select those disruptive technologies that can provide you maximum advantage in terms of business process optimization and the competitive advantage.

By now, you will have got a real feel of how different AI-based technologies are reshaping the future of maritime shipping. You might have asked a question to yourself, how the AI-based systems gained so much acceptance in all fields and industrial domains? Why more and more business entities are striving to implement AI-based systems. This question is nicely answered by Dr. Tony Bates in his book on digital age.clii In this book, he has mentioned four key benefits of implementing AI as shown in Figure 107 below:

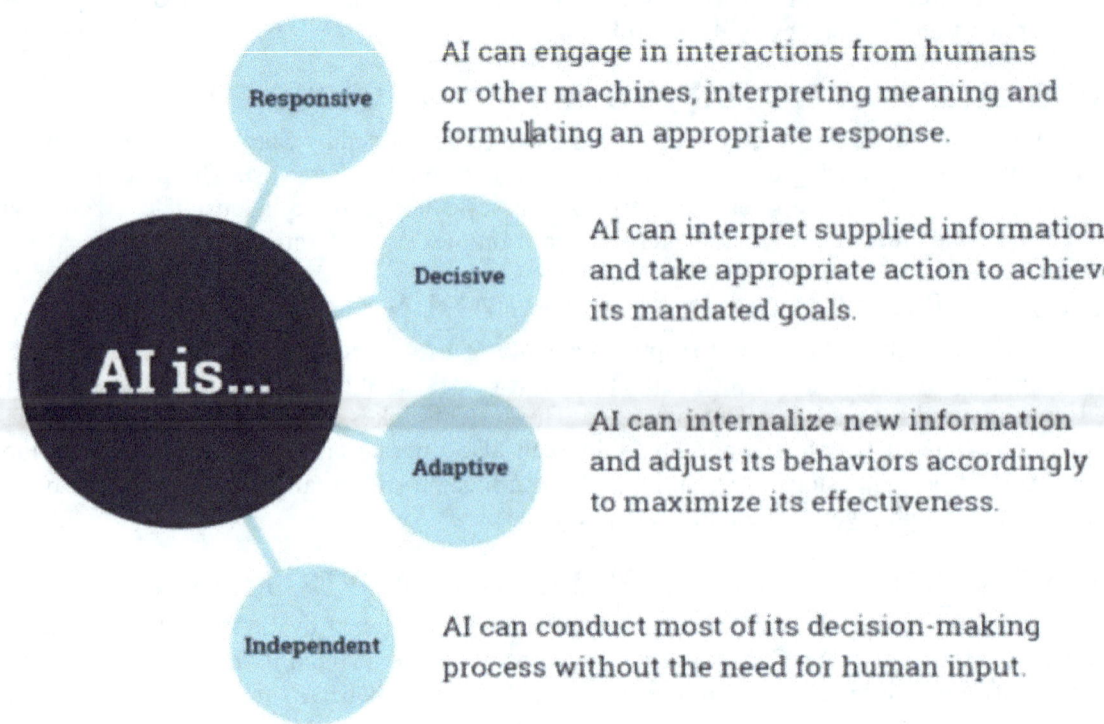

6.10. FIGURE 107: KEY BENEFITS OF IMPLEMENTING AI-BASED SYSTEMScliii

Figure 107 highlights that the first key benefit of implementing AI-based systems is that these systems are highly responsive and always formulate an appropriate response. The second benefit gained form AI-based implementations is the decisive nature of these systems. The supplied data can be interpreted by the AI algorithms that facilitates data-driven decision making. The third benefit is the adaptability of the AI-based systems. The systems adjust their recommendations, responses, and suggestions based on the new information and the AI model is optimized in a continued manner. It is the feature that is completely absent in the conventional computer applications. The conventional systems have no variations in their responses on day one and five years after their usage. The fourth benefit is the independent nature of the AI-based systems. These systems carry out the decision-making process with minimal or no human intervention. If the systems are designed to make corrective actions, they can even respond to emerging threats and challenges.

With all these benefits, AI-based systems should truly be a choice for the ship owners, ship managers, maritime institutes and organizations. In this chapter, I have presented comprehensive details on emerging technologies, predictive analytics tools and techniques, and disruptive technologies. Now, you have an extremely good level of understanding of the different variants and techniques of AI-based systems. This book will continue to be a valuable resource for you for optimizing your decisions in the maritime shipping organizations regarding AI-based implementations.

It was the last chapter in which I have highlighted different aspects and dimensions of unleashing AI for transforming maritime shipping for the future. The next chapter is the conclusion chapter. In the conclusion, I have summarized all the key ideas presented in this book. However, when you go for actual implementation in your organization, it is highly recommended that you read all the relevant sections of the book in detail besides reading the summarized key ideas. I have also presented the key takeaways from this book. You must have enjoyed reading each and every section of this book. I have done extensive research before writing all these sections. Although I possess valuable knowledge and skills on my own, but still I went through an extensive academic literature before completing this work. Your journey and reading the whole book is an endorsement for me that you appreciate the valuable content presented in this book. Let's move to the last chapter of this work and culminate the exciting journey of reading this book.

7. CONCLUSION

Ah! It's now time to write the conclusion of this book. It is giving me a sense of satisfaction and accomplishment that my efforts and research work got your attention and you have read all the sections of the book and reached here. However, the sad part is that it will end my relationship and bonding with you when you are done reading this book. Nevertheless, I feel that this book is a source of continued guidance and inspiration for the ship owners, ship managers, maritime institutes and organizations. So, you will keep on referring this book during the course of your AI-based implementations. I will be highly pleased if you share your success stories concerning AI-based implementations in your organization and how this book helped you in these endeavors. You can share your stories to me at my email mentioned in About the Author section.

This chapter is devoted to summarizing the key ideas of the book. I also mentioned earlier that some readers tend to read the summaries during the implementation phase. However, it is always a good strategy if you read that entire section that is applicable in your implementation scenario. Let's move to the different sections of the conclusion chapter.

7.1. SUMMARY OF THE KEY IDEAS

I have given you comprehensive details right from the basics of AI to the different applications of AI in maritime shipping. It is not possible for me to summarize everything I covered in just one or two paragraphs. Therefore, I have also allocated a sufficient portion of this book for summarizing the key ideas.

AI is the simulation of the processes of human intelligence. The computer systems, machines, and machine learning algorithms attempt to simulate the responses of human intelligence in given situations and scenarios. AI implementation requires a specialized IT infrastructure for writing the software code and training the algorithms of machine learning. AI program development is not proprietary to any single programming language. AI developers use different programming languages such as Python, Java, R, Julia, and C++. AI developers create computer algorithms and programs such that the computer systems get the ability of processing the large volume of information. The algorithms perform the tasks of learning from the datasets, reasoning, and rational decision-making. The ultimate objective of the AI developer is to simulate the cognitive functions of human brain and make adaptations in the decision-making approach similar to humans.

Many individuals confuse the term machine learning and use it interchangeably with AI. Machine learning is just one of the key components of AI, otherwise AI is highly diversified domain. Other crucial components of AI include neural network, deep learning, natural language processing, computer vision, and cognitive computing.

The world of today is a data-driven and fast-paced world. Therefore, the significance of AI should not be underestimated by the maritime industry professionals. Some maritime organizations have already embraced AI and it is eventually going to transform the maritime sector. AI will address the age-old challenges of maritime shipping and build a more sustainable, efficient, and competitive future. Many maritime institutes and organizations are reluctant to implementing AI because they feel that it will incur a significant initial cost. There is no denying of the fact that the development of the required IT infrastructure and the acquisition of computer systems with high processing power will incur a significant initial cost. However, ship owners, ship managers, maritime institutes, and organizations should have a long-term orientation. They should analyze how it will make the life easier and improve the accuracy and safety of the whole maritime shipping operations.

Different successful cases of AI-powered maritime transportation were presented in this book. Rolls-Royce (RR) has an accomplished business at the international level related to marine systems and it is a prominent manufacturer of marine systems and marine engines. RR capitalized on AI-powered systems and introduced an Intelligent Awareness System on vessels. The system was based on a connected network in which sensors and cameras were used by the AI algorithms. The programs process and interpret the received data and facilitate the crew members by presenting to them a report of situational awareness. It is a highly beneficial report for the staff because it helps them in detecting other vessels in the sea, identifying the potential obstacles, and getting navigation recommendations for the voyage safety. RR reported that this AI-based intervention improved the performance of the crew significantly and they were able to make intelligent decisions for reducing accidents and collisions. Maersk Line (ML) enjoys the leadership among the shipping companies and the company is also providing leadership in the deployment of AI-powered systems. The predictive maintenance is one area of maritime shipping where AI has been utilized successfully by ML. AI algorithms receive data from the sensors and perform a critical analysis of the current health of the ship. They predict the areas where the maintenance has become due. They also indicate the components and critical machinery that will be needed for carrying out the maintenance work.

CMA CGM (CC) is a well-known company in the domain of container shipping. The company made use of robotic systems powered by AI technologies. This system is being used by the container shipping company for automated handling of the cargo. The robots at the company are equipped with AI systems. They have also been trained by using the computer vision technology. They perform the tasks of loading and unloading the containers from the ship without human intervention. It has enabled the company to reduce the staff cost and handle the cargo at the terminals efficiently. The real benefit that CC got from this intervention was a reduced time to handle the cargo that provided them with a competitive advantage. Moreover, there was a higher level of efficiency in the operations of cargo handling.

Wartsila (Wart) is a provider of innovative technology solutions and the company has a proven track record in introducing technology-based marine solutions. The company developed a new product known as Smart Marine Ecosystem. This AI-powered system evaluates the data gathered from the vessel operations and uses it for the optimization of fuel consumption. The data is also used for recommending maintenance schedules and navigation routes. Kongsberg (KB) offers advanced IT solutions for the maritime shipping industry. KB developed a DP (Dynamic Positioning) system for the ships that was powered by AI technologies. The system capitalized on the data received from multiple sensors configured on the ship. The data processing by AI algorithms made it possible for the ship operators to maintain an accurate position of the vessel. It was particularly useful during challenging weather conditions when it was required to maintain the stability of the vessel. The risks in the critical vessel operations reduced significantly by this deployment and the operational efficiency was enhanced remarkably.

AI can be used in the maritime transportation for optimizing the vessel operations, predictive maintenance, cargo optimization, navigation and enhanced safety, fuel efficiency, route optimization, and preventing accidents and collisions.

The application of AI can also be seen as to how the AI-based systems influence and optimize the maritime business. Some of the maritime institutes and organizations have already embraced AI in maritime shipping and it has provided promising results to those organizations. Other maritime businesses will also have to embrace this technology because AI is not only 'enemy' of the conventional jobs but it will also eliminate those businesses that insist on doing businesses in the old fashion. The first successful case is of Nautilus Labs (NL). NL has optimized the fuel efficiency of the vessel voyage by using AI-based predictive analytics. This AI-based platform makes use of machine learning techniques for evaluating speed, weather conditions, and engine performance. The insights and recommendations by the AI algorithms are used for optimizing the fuel usage. It has resulted in a substantial cost reduction in NL and the company has also been able to improve its environmental track record. Another renowned vendor offering AI-powered solutions is Shone. The company offers AI-powered systems that optimize the navigation capabilities and the offered systems are also beneficial for autonomous ships. The technology used by Shone is based on sensor fusion, computer vision, and machine learning techniques that ensure the safety of vessel operations. When maritime institutes and organizations implemented AI-based systems offered by Shone, a higher level of autonomy was achieved in vessel operations and the organizations were able to reduce the count of onboard crew.

The third prominent vendor is Orolia where the company offers situational awareness solutions. Collisions in vessel occur when the area is congested and the ship captain has a reduced level of situational awareness. This deficiency can easily be overcome by deploying situational awareness solutions such as the one offered by Orolia. The system integrated AI algorithms with satellite data, radar, and AIS. The system detects the safety and security threats and facilitates the captains in making informed decisions. The maritime institutes and organizations that have implemented Orolia's situational awareness have been able to ensure the safety of the vessel voyage and handle the potential threats in the surroundings. Another notable vendor for maritime businesses is Nautix Technologies (NT). The system makes use of the predictive maintenance feature of the AI. The maintenance algorithms are applied on marine engines and the associated equipment. The system analyzes the historical data as well as the current data from the sensors. The maintenance actions are recommended by the system to avoid equipment failure. Many maritime institutes and organizations have installed NT systems and report significant reductions in downtime and maintenance costs. Therefore, this system should be used by the maritime organizations because it helps them in proactive maintenance planning. Another prominent vendor for maritime businesses is Voyager Analytics (VA). The company has added a new dimension to AI in the context of maritime shipping. The company offers those solutions that are responsible for the performance and welfare of the crew members. The capabilities of natural language processing are used for assessing the communication among crew members and providing them relevant feedback by the AI algorithms. The recommendations are aimed at improving the job satisfaction of the vessel crew and ensuring their mental health and wellbeing. Maritime companies implementing VA have reported higher job satisfaction of the crew members and the staff turnover rate is also very low in these organizations.

It is important to note that the maritime shipping companies will have to acquire solutions from one of the vendors that are offering solutions in their geographical locations. It is not possible for the maritime shipping companies to build AI-based systems on their own. These systems are based on highly sophisticated technologies that cannot be learned easily by maritime shipping companies. IT skillset is a completely different domain and these systems can best be developed by software houses. The maritime institutes and organizations should prefer the acquisition of those systems that are based on global standards and the best practices.

AI is also optimizing various aspects and factors of maritime logistics. Its key interventions can be seen in the effective inventory management and demand forecasting. The algorithms evaluate the historical data of sales and forecast the demand of maritime shipping with accuracy and precision. AI tools can also highlight market trends for intelligent decision-making. When the companies adopt this forecasting approach, maritime shipping companies can avoid stockouts and minimize carrying costs. Another key intervention of AI is in efficient delivery planning and route optimization. The scheduling and routing of vessel voyage are optimized by AI algorithms where the system considers the parameters of delivery window, vessel capacity, weather conditions and traffic situation.

The most critical factor in the supply chain of maritime shipping is the relationship optimization and supplier management. The AI algorithms evaluate the performance of a supplier based on the historical data and they consider the factors of costs, quality of service, and delivery times. As a result, the most reliable suppliers can be selected. If the organization wants, the other suppliers may also be communicated regarding their reasons of rejection by the AI algorithms so that they could improve their services for a future selection. AI-based systems also improve the collaboration and relationship with the suppliers and streamline the entire process of procurement.

AI can transform decision-making in maritime shipping in multiple areas and fronts. The first key area is the processing and analysis of data. The AI algorithms can process both structured data and unstructured data. You will be surprised by the unprecedented speed of AI systems in processing big data. The decision-makers then gain valuable insights and find the patterns in data that enable intelligent decision-making. Machine learning techniques form an integral component of the overall AI tools. These techniques uncover hidden patterns in the data and then predictive analytics can guide ship owners, ship managers, maritime institutes and organizations. Another important area of intervention is the forecasting tools. In maritime shipping, the environmental variables have a higher level of uncertainty and the captains have to make key decisions while the vessel is travelling in the sea. The predictive models of AI facilitate the tasks of ship owners and ship managers because the future trends are forecasted based on the analysis of historical data. These tools also assist in predicting customer behavior and enabling proactivity in the decision-making process.

When AI-powered systems are used, the surveillance and monitoring of the vessel voyage can be accomplished in real-time. These systems use computer vision to evaluate the video feeds of the vessel operations. The suspicious behavior and unauthorized accesses are immediately reported by the system. The threats are detected by the system by processing vast amounts of data. The systems not only focus on the sea operations but also evaluate public safety concerns and cyber threats. Machine learning techniques provide information regarding the suspicious network activity and data security of the connected environment is also ensured. Various touchpoints in a large vessel can also be controlled by AI-based systems of biometric recognition. The finger prints and facial patterns can easily be recognized by these systems for the identification of the crew members and access control. The access may be denied to certain privileged areas based on the access role matrix.

The planning and scheduling of the staff can also be optimized by AI algorithms. The systems analyze the skills level and preferences of the staff and assign them the duties accordingly. The algorithms also ensure the compliance of the HR business processes with the state and federal labor laws and regulations. When the AI-based systems are introduced in maritime institutes and organizations, there is a need for training the crew for using these new systems. The AI experts will just install and configure these systems and then it is up to the crew to make the most of these new systems. The systems are beneficial only up to the extent to which the intelligent reports are utilized by the crew and the management.

AI-powered systems evaluate the user behaviors in the interconnected systems and also assess the overall network traffic. The anomalies and unusual patterns in the network traffic are immediately reported to the data centers. The machine learning techniques continuously evolve them to combat the recent strategies of the hackers. The AI algorithms also detect if there is any malware injected in those systems that are using AI tools. The presence of malicious software, ransomware, and viruses all are detected by the algorithms. The threats are blocked before the hackers are successful in their motives. AI-based antivirus solutions are also being offered currently that makes it possible to fight the threats through the real-time protection.

Customers now have higher expectations from the maritime organizations and they want the availability of the customer support round the clock. Earlier, it was challenging for the organizations to enable customer interactions all the time. However, virtual assistants and chatbots have enabled the business entities to process a high volume of customer queries round the clock. The good aspect of these AI-based technologies is that they are expert in the processing of natural language such as the chatbots. The bots can understand the customer queries and concerns and respond in a way similar to humans. In this way, the response time improves significantly and the accuracy level of the provided information is also improved. In the case of a human assistant, the individual may forget a point and provide wrong information. However, chatbots respond based on the data fed to them. The humans may also get frustrated and angry with the aggressive remarks of the customer, whereas the chatbots listen to all the responses without anger. The customers may also find it convenient to talk to bots because they can talk with an open heart without a fear that a given comment will be disliked by the opposite party. AI-based systems and services also provide customized recommendations to the customers based on their past visits to the website and purchase history. The promotions and recommendations regarding the maritime shipping services are always based on the preferences of the visitor. This aspect of personalization improves customer satisfaction and customer engagement.

There are also various challenges associated with AI-based implementations and these challenges can broadly be classified into technical, regulatory, and cultural challenges. Strategic solutions are required for addressing the technical challenges in the maritime shipping because the challenges are intricate and some of these challenges are specific to maritime shipping. One of the topmost challenges that may be faced in the maritime shipping sector is the availability and quality of data. The AI algorithms are only as intelligent and competent as the training data available to them. If you cannot provide a good quality of data in a sufficient volume to the algorithms, the results produced by AI algorithms will not be promising. Another technical challenge that might be observed in maritime shipping is the level of standardization and integration achieved in the underlying data. Various systems and sensors provide data to maritime shipping systems that are powered by AI. The data might be collected from the weather stations, the vessel itself, the logistics chains, and the ports. It is a really complex and challenging task for the AI algorithms to standardize and integrate this data into AI systems. Another technical issue that may arise in the maritime shipping is in the context of connectivity and edge computing. Maritime vessel operations are all about operations in the far flung areas of the sea where there may be a low connectivity in a remote area. When the AI-powered systems are used, all the connectivity and interfacing among various systems is achieved by the internet cloud. The real-time transmission of data may become highly challenging and complex when a good level of internet connectivity is not available for transmitting the data to onshore AI systems.

The regulatory challenges in AI implementation also need to be examined carefully by ship owners and ship managers because they have far-reaching implications. They might even result in completely abandoning the AI-based system because the performance of the AI-based systems is reliant on the quality of the data and the availability of data. If the restrictions are imposed by the regulators on the availability of certain data points, the performance of the AI-based systems will degrade substantially. In the worst case scenario, these systems might work like conventional and ordinary systems in the absence of data because there is not a process of learning and adaptation.

The first and the biggest regulatory challenge in the AI-based implementations is the adherence of the maritime shipping companies with the maritime regulations. Different regulatory frameworks are applicable to maritime institutes and organizations. Another regulatory challenge that makes the AI implementation in the maritime shipping industry a daunting task is the lack of AI-based regulations. It keeps the senior management always worried that if there are new regulations introduced in the future, their whole investment might go wasted. There is one more challenge in the regulatory paradigm and it is related to security compliance and data privacy. When the AI algorithms use personal and sensitive information, compliance issues may emerge. Therefore, the ship owners and ship managers should also take a legal advice regarding the data privacy laws. In this regard, the internationally known regulatory framework is General Data Protection Regulation (GDPR).

AI implementation also poses cultural challenges for the maritime shipping organizations. It requires a significant business process reengineering and therefore, there should be a paradigm shift in the current organizational culture. The biggest challenge that the ship owners, ship managers, maritime institutes and organizations face is the resistance to change and a traditional mindset. The business processes of the maritime shipping are highly complex compared to the road shipment and air shipment. Another cultural challenge is to meet the training needs and skills gap of the current workforce. AI is all about the use of advanced, state-of-the-art technology and the end users should also have a good level of technological literacy. There is one more cultural challenge associated with AI implementation and it is the fear of losing the job. When the systems are automated, the current workforce will definitely have a realization that their services have become redundant. Think about autonomous ships, automated response systems, automated predictive maintenance, robots for container handling, robots for cargo handling, the fear factor in these circumstances is only natural. The ship owners, ship managers, maritime institutes and organizations should address these concerns of the current workforce.

MDPI, in a journal article, has presented a roadmap for an effective AI implementation in four key steps. The first step is to gain a good level of understanding of AI tools and technologies and then assess the level of implementation needed in a maritime shipping company based on the organizational capabilities. The second step is to review the current business model (BM) of the organization and identify the areas where the AI systems should be implemented and new business model implementation (BMI) is required. The third step is to make the current systems capable and ready for the AI migration. The fourth and final step before the AI implementation is to gain an acceptance level of the AI implementation in the maritime shipping organization.

There is also a six steps process of implementing AI data model. The first step is to develop a good level of understanding of the business. The maritime shipping business professionals already possess a good level of the understanding of their business domains, but they will need to view the business from the perspective of AI-based systems. The second phase is the understanding of the data. The ship owners and ship managers should focus on the readiness of the data for the system migration and also ensure the full transparency concerning the acquisition of data that may also include the receiving of an informed consent from the relevant parties. The third step is the preparation of the data. At the end of the day, the AI-based systems also save the data in the databases such as oracle and sql server. The tables and fields are defined in those database platforms. The data from the existing systems should be prepared in a format that is consistent with the new systems and can be easily imported into the new AI-based systems. The fourth and the most important phase is data modeling. The ship owners and ship managers can use a proven and tested approach for data modeling. The fifth phase is the evaluation of different use cases that have been developed by the AI programmers. The execution of the use cases will make the ship owners and ship managers a real feel related to the significance of using AI-based systems. The sixth, and the final phase, is the deployment of the AI-based solution. The deployment should be made after an extensive testing of the system. The system should be tested based on the unit testing, the whole system testing, as well as the integration testing.

Risks may be experienced in AI-based implementations from multiple dimensions. These dimensions together create an identity risk for the maritime shipping organization. Therefore, the ship owners, ship managers, maritime institutes and organizations should mitigate the risks associated with all these dimensions for a successful AI implementation. The first dimension is known as the strategy dimension. The fate of an AI project may also go wrong even at the strategy level. The second dimension is called trust dimension. The functionality of an AI-based system resembles a black box where the end users have a little idea regarding the working of the algorithms. The user experience might be poor if they are not well trained in using the AI-based systems effectively. The third dimension is the dimension of ethics. The dataset fed to the AI-based algorithms may not represent the whole population. In those cases, the learning of the AI algorithms will be wrong and it will also reflect in their decision-making approaches. The system may also suffer from the algorithmic bias and the data bias. The fourth dimension is known as the compliance dimension. The data acquisition may violate the data privacy laws of various countries and regions. The regulatory oversight may also get reduced due to the higher level of automation. The fifth dimension is the financial dimension. AI-based systems and infrastructure development are highly expensive. If there are delays in the data acquisition and the development of AI model, then cost may increase even more. The sixth dimension is known as the technology dimension. The technical skills of the IT and the functional team in the maritime shipping organization may not be up to the mark for the AI implementation. The error analysis may also be weak. The seventh and the last dimension is the people dimension. If there are extreme weather conditions, the AI algorithms might have to be reworked to calculate all parameters that will increase the workload of the crew members. The organization may also face resistance from environmental protection organizations for implementing technology-intensive projects in the maritime region.

The ship owners, ship managers, maritime institutes and organizations should also be well aware of the ethical implications of their initiative or else they will have to face penalties and fines not only from the government agencies and the regulatory bodies, but the customers may also lodge complaints against those companies.

The first type of ethical implications may be related to the international regulatory framework. It is because the maritime shipping operations are transnational operations. There are various guidelines already developed in the international regulatory frameworks for a responsible use of AI.

The first and primary source of ethical premise is United Nations Guidelines. The UN has made extensive efforts for developing a comprehensive ethical framework so that all member states follow some standards while implementing AI. The guidelines of the UN were known to the general public in 2021. These guidelines highlight that the basic human rights and ethical considerations should be adhered to in all AI-based implementations. The implementations should not harm the interests of the stakeholders or the surrounding communities. The second guiding framework for ethical considerations is OECD Principles. The Organisation for Economic Co-operation and Development (OECD) has also followed the footsteps of the UN and developed its own ethical guidelines for implementing AI systems. The guidelines emphasize that the business managers should make a responsible use of AI. It means that the AI should be utilized in the maritime shipping organizations by considering the factors of transparency, fairness, and accountability. The third important resource on ethical guidelines is the UNESCO Recommendations. The United Nations Educational, Scientific and Cultural Organization (UNESCO) has also realized that the AI-based implementations come under the purview of scientific and educational initiatives, and therefore, ethical guidelines have been developed. These guidelines mention that the implementations should follow the principles of autonomy, equity, and accountability.

When there are refinements in different AI models and there is a wide scale implementation of AI in the maritime shipping industry, there will be temptations among the ship owners, ship managers, maritime institutes and organizations to spread the use of AI in all business processes and systems. The use of AI should be within the ethical boundaries and comply with the international and the local regulatory framework. This notion is now popularly known as the 'responsible use of AI'.

The analysis of the environmental impact also comes under the purview of the ethical considerations. It is because if the maritime shipping companies are involved in embracing AI for gaining a competitive advantage, they cannot do so at the cost of harming the environment and consuming more energy resources.

Another area of consideration is an Eco-Friendly Data Management and the Green Data Centers. The maritime shipping companies should implement sustainable data practices. The ship owners, ship managers, maritime institutes and organizations should use the techniques of data compression, efficient data storage, and data center location choices that are aligned with the sustainability agenda. They should reduce the environmental impact of data management in AI based systems. The ship owners, ship managers, maritime institutes and organizations should build partnerships with green data centers that give preference to sustainability. These data centers utilize energy-efficient systems, renewable energy sources, and cooling systems to reduce the carbon emissions related to AI based implementations.

The cultural challenges associated with AI implementation are that the AI based implementations also create a sense of job insecurity among the population. The crew members tend to believe that AI implementations will eat their jobs and more and more robots will replace the humans. The senior management in those organizations should make the workers realize that AI implementations are for improving the quality of output and enhancing the efficiency. The knowledge and experience of the hired workers will still be respected and the implementations will not result in downsizing or retrenchment. There are five key examples of the startups that have shown an impressive growth in their businesses by offering maritime AI solutions based on machine learning technologies. The purpose of highlighting these five examples was to make you realize that the proponents of AI still have a very optimistic view regarding the benefits of using AI-based systems. An advanced learning of the usage and development of these systems have also enabled the individuals to build their own startups and offer their customized solutions based on their skills and knowledge. Therefore, the ship owners, ship managers, maritime institutes and organizations should also be hopeful regarding the potential of AI in transforming maritime shipping for the future.

Artificial intelligence is regarded as one of the frontier technologies. The blockchain based systems and quantum computing systems also implement the concepts of artificial intelligence. A new technology paradigm in maritime AI has been introduced by MDPI in a journal article. By following the analogy of Internet of Things (IoT), this paradigm has been termed as Internet of Underwater Things (IoUT). This framework is highly significant for the ship owners, ship managers, maritime institutes and organizations because internet connectivity in the maritime region will pose new challenges and issues when AI-based systems will be implemented in the maritime shipping. The ship owners will have to enable connectivity of the ship through the satellite. This connectivity can be made possible by electromagnetic communication. The monitoring center will also receive data via satellite. The offshore data center should also be connected with the satellite so that there are no wired connections affecting the receiving of high-speed internet bandwidth.

Maritime shipping is a combination of the maritime operations and the shipment operations. There are also various logistics arrangements involved outside the sea area. Therefore, the ship owners, ship managers, maritime institutes and organizations can realize the full benefits of AI if they focus on its implementation not only from the perspective of maritime AI but also form the lens of shipment and logistics AI.

If the data to be processed is very small, then the ship owners and ship managers can even be comfortable with using the conventional computer systems and applications. If there is a slight growth in the database, then one can use descriptive analytics systems that will tell what happened at different touchpoints of maritime shipping and you will get a holistic picture through these systems. If you are interested in gaining more sense of your data and also want to know the causes of certain system behaviors, then you should go for diagnostic analytics systems. Predictive analytics is a stage next to diagnostic analytics. It is a stage where a maritime shipping company has implemented AI tools and technologies. As a result, the AI algorithms and machine learning technologies are processing huge datasets from various systems and interfaces. The use of such heavy processing should result in meaningful insights for the business entity. Therefore, in the next stage, you will be required to install predictive analytics systems. These systems will highlight to the ship owners, ship managers, maritime institutes and organizations what is going to happen during the vessel voyage and shipment operations. These key insights will be useful for both the business managers and the customers.

In the domain of information and communication technology, those technologies are considered as disruptive technologies in which innovation is at a highly advanced level such that there is a paradigm shift in the way businesses, consumers, and industries are operating at present. From the ancient ages to the present, wheel, the bulb, and the mobile phones can be categorized as disruptive technologies because they entirely changed the way businesses are being conducted globally. The ASEAN Post has mentioned the list of 12 disruptive technologies, and artificial intelligence is at the top of the list. Its key relationship is with the automation in the knowledge society. AI also has relationship and dependence on other disruptive technologies such as global internet penetration, Internet of Things, cloud technology, advanced robotics, and augmented reality.

The above description summarizes various ideas and concepts presented in this book. I have tried to produce a summary that could cover all the aspects and dimensions. However, as I explained to you earlier, you will definitely have to go into details at the time of actual implementation process.

7.2. KEY TAKEAWAYS

There are several key takeaways for the readers of this book. My intent of writing this book was to make the ship owners, ship managers, maritime institutes and organizations realize the true potential of AI for transforming maritime shipping for the future. I have elaborated the significance of AI tools and technologies through various dimensions in this book. You now have a good level of understanding of the theoretical framework of AI and how these theories have been put into workable solutions in the maritime shipping industry. You are also aware of the benefits of AI in maritime transportation. You know various successful implementations of AI-based solutions and are also aware of the key vendors and solutions for the AI-based systems in the maritime shipping.

This book has also equipped you with the knowledge of the significance of AI in the maritime business. You know how AI can optimize logistics and the entire supply chain. You are well aware how AI systems and tools can be used for intelligent decision-making, optimized safety and security, and an effective crew management. You also possess the knowledge regarding AI for crew management and AI for a better customer service.

You have also been given a good level of understanding regarding the technical, regulatory, and cultural challenges that might be faced during the AI implementation. As the ship owners, ship managers, maritime institutes and organizations, you will also be required to adhere to the ethical guidelines and the responsible use of AI and all these ethical considerations are now known to you. You have also come to know regarding the emerging AI-based technologies, disruptive technologies, and predictive analytics systems.

7.3. CALL TO ACTION FOR SHIP OWNERS, SHIP MANAGERS, MARITIME INSTITUTES AND ORGANIZATIONS

It's now time to get the full benefit of this book by using the key takeaways for implementing AI-based solutions in the maritime shipping companies. I urge all the ship owners, ship managers, maritime institutes and organizations to realize the significance of unleashing AI for transforming maritime shipping for the future before it is too late and the competitors lead these AI-based initiatives. If you lead the AI-based ventures in the maritime shipping industry, your performance will be substantially better than your competitors, and you will achieve a long-term sustainability of your business.

7.4. RECOMMENDATIONS FOR THE SUSTAINABLE OPERATIONS OF MARITIME SHIPPING

I have written this book because I possess a huge experience of the maritime shipping industry and I know that the industry is at a critical juncture of the history. It is a make or break point for the industry and the ship owners, ship managers, maritime institutes and organizations should consider the sustainable business operations as a first priority.

The ship owners and ship managers should realize that the initial investment associated with the AI-based infrastructure, systems, and applications will be recovered in the short run. They should be more worried about the sustainability of the maritime shipping business in the future when the other organizations would have optimized, AI-based maritime shipping processes. They will really struggle to survive in those times. Therefore, the time to react is just now. Unleash the potential of AI and implement AI in maritime transportation, shipment and delivery, crew management, logistics, and other areas of the maritime shipping.

It is a good time to embrace AI because you will be counted among those few who have implemented AI in all areas of maritime shipping. You will be regarded as pioneers in technology adoption. As I also gave to you examples of successful machine learning technology startups, you can also build upon your expertise and become a good entrepreneur in AI-based technologies once a few of your AI-based implementations in the maritime shipping companies are completed smoothly. As I explained to you, the implementation process is not a bed of roses, and you will face various technical, regulatory, and cultural challenges. You will also be required to follow the ethical guidelines during implementation. But sooner or later, you will have to start choosing the pathway of AI-based implementation of systems. So, why not start it now? It is really a time to rise to the occasion.

8. ABOUT THE AUTHOR

The Author is a maritime visionary with a captain's heart and an island soul. The author has used all his knowledge, experience, and skill set, acquired by working in the maritime shipping sector, for explaining different dimensions of maritime shipping AI in this book. In his island home, the sea's love, sailing's legacy, and leadership's flame passed down through generations with pride and glory. He is a skilled navigator of words, charting a course through the vast ocean of knowledge. He believes in every word and section written in this book and all the content is the outcome of a critical analysis, intellectual discourses, and key debates concerning maritime shipping AI. With his expertise and passion, he guides readers towards prosperous shores, unveiling the secrets of maritime life and business success in concise and captivating prose.

REFERENCES

i. https://www.techtarget.com/searchenterpriseai/definition/AI-Artificial-Intelligence

ii. Jarrahi, M. H. (2018). Artificial intelligence and the future of work: Human-AI symbiosis in organizational decision making. *Business Horizons, 61*(4), 577-586.

iii. Johansson-Pajala, R. M., Thommes, K., Hoppe, J. A., Tuisku, O., Hennala, L., Pekkarinen, S., & Gustafsson, C. (2020). Care robot orientation: what, who and how? Potential users' perceptions. *International Journal of Social Robotics, 12*(3), 1103-1117.

iv. Manikonda, L., Deotale, A., & Kambhampati, S. (2018, December). What's up with privacy? User preferences and privacy concerns in intelligent personal assistants. In *Proceedings of the 2018 AAAI/ACM Conference on AI, Ethics, and Society* (pp. 229-235).

v. https://www.spiceworks.com/tech/artificial-intelligence/articles/what-is-ai

vi. Ghobakhloo, M. (2020). Industry 4.0, digitization, and opportunities for sustainability. *Journal of Cleaner Production, 252*, 119869.

vii. https://www.britannica.com/technology/artificial-intelligence/The-Turing-test

viii. Ni, J., Chen, Y., Chen, Y., Zhu, J., Ali, D., & Cao, W. (2020). A survey on theories and applications for self-driving cars based on deep learning methods. *Applied Sciences, 10*(8), 2749.

ix. https://www.ibm.com/topics/industry-4-0

x. https://www.spectralengines.com/articles/industry-4-0-and-how-smart-sensors-make-the-difference

xi. Lv, Z., Wu, J., Li, Y., & Song, H. (2022). Cross-layer optimization for industrial Internet of Things in real scene digital twins. *IEEE Internet of Things Journal, 9*(17), 15618-15629.

xii. https://industrywired.com/artificial-intelligence-in-shipping-is-stepping-into-new-frontier

xiii. Octavian, A., & Jatmiko, W. (2020, October). Designing intelligent coastal surveillance based on big maritime data. In 2020 International Workshop on Big Data and Information Security (IWBIS) (pp. 1-8). IEEE.

xiv. Nishant, R., Kennedy, M., & Corbett, J. (2020). Artificial intelligence for sustainability: Challenges, opportunities, and a research agenda. *International Journal of Information Management, 53*, 102104.

xv. https://nexocode.com/blog/posts/ai-in-maritime-artificial-intelligence-solutions-in-the-shipping-sector

xvi. Himeur, Y., Elnour, M., Fadli, F., Meskin, N., Petri, I., Rezgui, Y., & Amira, A. (2023). AI-big data analytics for building automation and management systems: a survey, actual challenges and future perspectives. *Artificial Intelligence Review, 56*(6), 4929-5021.

xvii. Thombre, S., Zhao, Z., Ramm-Schmidt, H., García, J. M. V., Malkamäki, T., Nikolskiy, S., & Lehtola, V. V. (2020). Sensors and AI techniques for situational awareness in autonomous ships: A review. *IEEE transactions on intelligent transportation systems, 23*(1), 64-83.

xviii. https://www.researchgate.net/figure/Conceptual-diagram-of-Smart-Autonomous-Ship-and-shore-service_fig1_326703030

xix. https://www.mdpi.com/2076-3417/10/17/6010

xx. https://www.gihub.org/infrastructure-technology-use-cases/case-studies/automated-robot-cranes-for-safer-ports

xxi. https://www.ediweekly.com/the-three-different-types-of-artificial-intelligence-ani-agi-and-asi

xxii. Ibid.

xxiii. Ibid.

xxiv. https://jashrathod.github.io/2020-11-05-basic-understanding-of-types-of-artificial-

	intelligence-for-providing-ingenious-solutions
xxv.	https://marine-digital.com/article_bigdata_in_maritime
xxvi.	https://smartshiphub.com/highlights/artificial-intelligence-to-monitor-vessel-performance
xxvii.	https://www.themanufacturer.com/articles/partnership-sees-rolls-royce-develop-intelligent-shipping-systems
xxviii.	https://d3.harvard.edu/platform-rctom/submission/ahoy-maersk-embraces-the-internet-of-things
xxix.	https://www.seatrade-maritime.com/asia/cma-cgm-deploys-innovative-container-tracking-solution
xxx.	https://www.wartsila.com/media/news/24-11-2017-wartsila-introduces-its-smart-marine-ecosystem-vision
xxxi.	https://www.kongsberg.com/maritime/products/positioning-and-manoeuvring/dynamic-positioning
xxxii.	https://www.hitachi.com/rev/archive/2023/r2023_01/01a06/index.html
xxxiii.	https://www.offshore-energy.biz/mol-launches-ai-powered-vessel-allocation-support-system
xxxiv.	https://www.awake.ai/post/ai-for-smart-ports-port-call-prediction-0
xxxv.	https://newatlas.com/robotics/ibm-autonomous-mayflower-ship-sets-sail-across-atlantic
xxxvi.	https://www.kaikosystems.com/blog/computer-vision-in-vessel-health-management
xxxvii.	https://www.youtube.com/watch?v=2V38iVkQ0r4
xxxviii.	https://www.mdpi.com/2077-1312/10/2/215
xxxix.	https://transporteca.co.uk/shipping-process
xl.	https://odsc.medium.com/ai-as-the-ultimate-disrupter-in-logistics-how-to-manage-last-mile-costs-c4874e8f2ea0
xli.	https://www.mol.co.jp/en/sustainability/innovation/case/safety
xlii.	https://www.mol.co.jp/en/sustainability/innovation/case/safety
xliii.	https://www.ship-technology.com/features/ship-navigation-system
xliv.	https://marine-digital.com/article_optimizing_vessels_route
xlv.	https://forwardermagazine.com/yara-marine-launches-route-pilot-ai-to-optimize-voyage-efficiency
xlvi.	https://sinay.ai/en/vessel-route-forecast-predicting-eta-and-increasing-fuel-efficiency
xlvii.	https://safety4sea.com/ai-can-help-improve-marine-collision-avoidance
xlviii.	https://www.inceptivemind.com/orca-ai-nyk-trials-new-safety-support-systems-autonomous-ships/21130
xlix.	https://www.fujitsu.com/global/about/resources/news/press-releases/2020/0415-01.html
l.	https://nexocode.com/blog/posts/ai-in-maritime-artificial-intelligence-solutions-in-the-shipping-sector
li.	https://www.adv-polymer.com/blog/artificial-intelligence-in-shipping
lii.	https://www.linkedin.com/pulse/how-generative-ai-revolutionizing-maritime-industry-justin-c-
liii.	https://www.linkedin.com/company/greywing
liv.	https://grey-wing.com/sea-gpt
lv.	https://grey-wing.com/sea-gpt
lvi.	https://www.ship-technology.com/buyers-guide/ai-companies-shipping-industry
lvii.	https://www.linkedin.com/pulse/navigating-future-shipping-analysis-artificial-its-impact-stefas
lviii.	Ibid
lix.	Ibid
lx.	Ibid
lxi.	Ibid
lxii.	https://www.freightwaves.com/news/qa-nautilus-labs-on-how-tech-can-tackle-fuel-use-and-emissions
lxiii.	https://aiforgood.itu.int/autonomous-shipping-is-making-waves

lxiv.	https://www.youtube.com/watch?v=lF_fPzgS_qM
lxv.	https://www.linkedin.com/company/nautix-technologies/posts/?feedView=all
lxvi.	https://www.voyager-labs.com
lxvii.	https://cyprusshippingnews.com/2022/02/23/windward-launches-ai-powered-ocean-freight-visibility-solution-to-tackle-supply-chain-visibility-challenges
lxviii.	https://windward.ai/solutions/ocean-freight-visibility-ofv
lxix.	https://windward.ai
lxx.	https://www.marinetraffic.com/en/online-services/plans
lxxi.	https://spire.com
lxxii.	https://www.sea-kit.com
lxxiii.	https://www.rivieramm.com/news-content-hub/news-content-hub/the-role-of-ai-in-voyage-optimisation-and-offshore-logistics-76760
lxxiv.	https://bdtechtalks.com/2021/05/20/ai-automated-shipping-logistics
lxxv.	Galaz, V., Centeno, M. A., Callahan, P. W., Causevic, A., Patterson, T., Brass, I., & Levy, K. (2021). Artificial intelligence, systemic risks, and sustainability. Technology in Society, 67, 101741.
lxxvi.	Angelov, P. P., Soares, E. A., Jiang, R., Arnold, N. I., & Atkinson, P. M. (2021). Explainable artificial intelligence: an analytical review. Wiley Interdisciplinary Reviews: Data Mining and Knowledge Discovery, 11(5), e1424.
lxxvii.	Alikhademi, K., Drobina, E., Prioleau, D., Richardson, B., Purves, D., & Gilbert, J. E. (2022). A review of predictive policing from the perspective of fairness. Artificial Intelligence and Law, 1-17.
lxxviii.	https://www.oceannews.com/news/defense/artificial-intelligence-what-are-the-challenges-of-securing-the-maritime-commons
lxxix.	Bharadiya, J. P. (2023). A Comparative Study of Business Intelligence and Artificial Intelligence with Big Data Analytics. American Journal of Artificial Intelligence, 7(1), 24.
lxxx.	Fotteler, M. L., Andrioti Bygvraa, D., & Jensen, O. C. (2020). The impact of the Maritime Labor Convention on seafarers' working and living conditions: an analysis of port state control statistics. BMC public health, 20(1), 1-9.
lxxxi.	Gyamfi, E., Ansere, J. A., Kamal, M., Tariq, M., & Jurcut, A. (2022). An adaptive network security system for iot-enabled maritime transportation. IEEE Transactions on Intelligent Transportation Systems, 24(2), 2538-2547.
lxxxii.	Chakriswaran, P., Vincent, D. R., Srinivasan, K., Sharma, V., Chang, C. Y., & Reina, D. G. (2019). Emotion AI-driven sentiment analysis: A survey, future research directions, and open issues. Applied Sciences, 9(24), 5462.
lxxxiii.	https://www.researchgate.net/figure/Connectivity-challenges-of-an-autonomous-ship_fig1_319900769
lxxxiv.	https://www.mdpi.com/2571-5577/6/3/58
lxxxv.	https://pub.towardsai.net/how-a-i-mistakes-can-have-real-life-impacts-e0a796e7a030
lxxxvi.	https://apacentrepreneur.com/the-three-major-limitations-of-ai
lxxxvii.	Ibid
lxxxviii.	Ibid
lxxxix.	https://ts2.shop/en/posts/experts-warn-of-dangers-as-shipping-adopts-ai-systems
xc.	https://www.imo.org
xci.	Ibid
xcii.	https://gdpr-info.eu
xciii.	Ibid
xciv.	https://transportgeography.org/contents/chapter7/transborder-crossborder-transportation
xcv.	https://ts2.shop/en/posts/experts-warn-of-dangers-as-shipping-adopts-ai-systems
xcvi.	Ibid.
xcvii.	https://ifa-forwarding.net/blog/european-logistics/the-shipping-industry-is-lagging-behind-on-embracing-ai
xcviii.	Ibid
xcix.	https://www.mdpi.com/2673-2688/1/2/11

c.	https://appinventiv.com/blog/ai-in-supply-chain-analytics
ci.	Ibid
cii.	Ibid
ciii.	https://appinventiv.com/blog/ai-in-supply-chain-analytics
civ.	https://www.bloomberg.com/news/newsletters/2020-08-12/supply-chains-latest-soaring-shipping-costs-set-to-go-higher
cv.	https://appinventiv.com/blog/ai-in-supply-chain-analytics
cvi.	https://www.globema.com/4-steps-ai-implementation
cvii.	https://www.sciencedirect.com/science/article/pii/S0925753520303052
cviii.	https://hbr.org/sponsored/2021/12/how-organizations-can-mitigate-the-risks-of-ai
cix.	https://towardsdatascience.com/7-types-of-ai-risk-and-how-to-mitigate-their-impact-36c086bfd732
cx.	https://www.arcweb.com/blog/mitigating-cyber-risk-maritime-applications
cxi.	https://www.nature.com/articles/s44221-023-00069-6
cxii.	https://emerj.com/ai-executive-guides/ai-advantages-challenges-coronavirus
cxiii.	https://www.wolterskluwer.com/en/solutions/enablon/bowtie/expert-insights/barrier-based-risk-management-knowledge-base/the-bowtie-method
cxiv.	https://www.mdpi.com/2071-1050/11/16/4501
cxv.	https://www.valuecoders.com/blog/blockchain-ml/unlocking-the-power-of-ai-real-world-examples-of-business-success
cxvi.	https://gestaltit.com/sponsored/intel/intel-2021/jimthewhyguy/ai-projects-fail-all-too-often-successful-ones-share-a-common-secret
cxvii.	https://www.talentica.com/blogs/implementation-of-ai
cxviii.	https://www.mdpi.com/2504-4990/3/1/4
cxix.	https://medium.com/marionete/ai-projects-8-factors-for-success-144ca6eb29aa
cxx.	https://www.dataiku.com/stories/detail/roi
cxxi.	https://www.orientsoftware.com/blog/ethics-in-ai
cxxii.	https://openeducationalberta.ca/educationaltechnologyethics/chapter/ethical-considerations-when-using-artificial-intelligence-based-assistive-technologies-in-education
cxxiii.	https://www.coe.int/en/web/bioethics/common-ethical-challenges-in-ai
cxxiv.	https://www.unesco.org/en/artificial-intelligence/recommendation-ethics
cxxv.	http://www.innblue.com.ar/News/News/201805/804349_169683_0.html
cxxvi.	https://www.startus-insights.com/innovators-guide/machine-learning-startups-maritime
cxxvii.	Ibid.
cxxviii.	Ibid.
cxxix.	Ibid
cxxx.	Ibid
cxxxi.	https://www.transformbase.com/tbaseblog/5-reasons-you-should-be-excited-about-the-big-five-frontier-technologies
cxxxii.	https://www.startus-insights.com/innovators-guide/maritime-trends-innovations
cxxxiii.	Ibid
cxxxiv.	https://www.mdpi.com/2077-1312/11/1/124
cxxxv.	https://marine-digital.com/article_5benefits_of_digital_twin
cxxxvi.	https://research.aimultiple.com/logistics-ai
cxxxvii.	https://www.frontiersin.org/articles/10.3389/fmars.2022.918104/full
cxxxviii.	https://www.heavy.ai/technical-glossary/predictive-analytics
cxxxix.	https://hgs.cx/digital/analytics/predictive-analytics-and-cognitive-analytics
cxl.	https://www.process.st/what-is-predictive-analytics
cxli.	https://www.csat.ai/how-artificial-intelligence-and-predictive-analytics-can-help-you-reduce-customer-churn
cxlii.	https://www.coppertreeanalytics.com/fundamental-series-on-building-analytics-artificial-intelligence-machine-learning-predictive-analytics-deep-learning-whats-the-difference
cxliii.	https://www.sv-europe.com/blog/an-overview-of-the-four-main-approaches-to-

	predictive-analytics
cxliv.	https://www.datasciencecentral.com/predictive-analytics-in-one-picture
cxlv.	https://theaseanpost.com/article/how-disruptive-technologies-are-transforming-southeast-asia
cxlvi.	https://www.mdpi.com/2673-8732/2/1/9
cxlvii.	https://www.startus-insights.com/innovators-guide/ai-solutions-impacting-maritime
cxlviii.	Ibid.
cxlix.	Ibid.
cl.	Ibid.
cli.	Ibid.
clii.	https://tell.colvee.org/mod/book/view.php?id=650&chapterid=1195
cliii.	Ibid.